T0181461

Design for Durability and Performance Density

Hani Ali Arafa

Design for Durability and Performance Density

Hani Ali Arafa
Mechanical Design and Production
Engineering Department
Cairo University
Giza, Egypt

ISBN 978-3-030-56818-4 ISBN 978-3-030-56816-0 (eBook)
https://doi.org/10.1007/978-3-030-56816-0

This Springer imprint is published by the registered company Springer Nature Switzerland AG
The registered company address is: Gewerbestrasse 11, 6330 Cham, Switzerland

Preface

A mechanical piece of equipment of a given design and size should satisfactorily endure its intended operation for a period of time that becomes longer under smaller loads and speeds, by the nature of things. Therefore, it would seem that the attributes of durability and performance density are conflicting or competitive design objectives. It might even seem that they are mutually exclusive ones.

This book explores the endeavors of the mechanical engineering community over a century or so, regarding the design for durability and the design for performance density, which endeavors have mostly been made without such connotations. It contains suggested as well as real-life examples of application of the principles and practices of mechanical design towards achieving these goals hence also examples of the so-called design pitfalls. Some of these examples might be regarded as ad hoc cases, by necessity, but they may enhance *design thinking* along similar lines. The book includes four case studies that reflect the author's contribution to solving problems in the manufacturing industry and in the oil field, of which failure event reconstruction invariably prove them to have a root cause in violation of, or non-abiding by the principles and practices of design for durability; they exemplify how engineers would design contrivances just to take their owners and operators later by surprise. The book also addresses several information voids that need to be filled, as well as problem areas that have been kept unspoken of, for long.

This book may but should not necessarily be regarded as fitting into the wealth of technical writings on (mechanical) Design for X, where the X stands for many different attributes that, however, do not hitherto include *durability* or *performance density*. This book aims to guide mechanical design and application engineers, as well as senior mechanical engineering students to

- Appreciate that design for durability does not essentially compromise performance density, and vice versa.
- Make a critical assessment of a given mechanical design regarding durability, performance density, or both.
- Avoid running into the various subtle design pitfalls and confronting their weird encounters, telling how things could be designed improperly or even the wrong way; to show how old experience could sometimes prevent novice mistakes.

- Save those young engineers from waiting for a lifetime to discover the bits of knowledge and experience that would have benefited them earlier on in their career.

Giza, Egypt Hani Ali Arafa

Contents

1 Introductory Definitions 1
1.1 Design for Durability 1
1.2 Loads and Loading in Mobile Assemblies 2
1.2.1 Determinate Loads 2
1.2.2 Indeterminate Loads 3
1.3 Design Pitfall Consequences; Weird Encounters 3
1.4 Performance Density 4
1.5 Design for Performance Density 5
References 6

2 Managing Reactive Loads 7
2.1 Reducing the Bearing Load by Proper Placement 7
2.1.1 Proper Location of an Idler Gear 8
2.1.2 Countershafts in Split-Torque Transmissions 9
2.1.3 Anomalous Gear Reducers 9
2.1.4 Case Study 11
2.1.5 Disc Brake Caliper Location 15
2.1.6 Worked Example 16
2.2 Design for Balanced Reactions 17
2.2.1 Balancing the Reaction Moments 18
2.2.2 Rolling-Mill Pinion Stands 19
2.2.3 Design Pitfalls with the Reaction Moment 20
2.3 Principle of Minimum Pressure Angle 22
2.3.1 Gearing Versus Traction Drives 22
2.3.2 Design for Reduced Side Thrust 22
2.3.3 Zero Pressure-Angle Engine Mechanism 23
References 24

3 Mutually Counteractive Loads at High Speed 25
3.1 The Simple Planetary Gear Set 25
3.1.1 High Transmission-Ratio Planetary Gearing 26
3.1.2 Demerits of Power-Recirculating Planetary Gearing 27

3.2 Coupled Planetary Gearing Characteristics 29
3.2.1 Generic Problem 30
3.2.2 Gearing with Intermeshing Planet Pairs 32
3.3 Sliding Between Spur Gear Tooth Flanks 34
3.3.1 Frictional Power Loss in Spur Gears 36
3.3.2 Alternative Estimation of Spur Gear Efficiency 37
3.3.3 Worked Example 37
3.4 Compound Planetary Gearing of High Ratio 40
3.4.1 Sunless Planetary Gearing 41
3.4.2 Sunless Planetary Twinset Gearing 43
3.4.3 Worked Example 44
References 46

4 Subtle and Obscure Loading Sources 47
4.1 Double-Helical Pinion Shuttling 47
4.1.1 Relative Positions of Double-Helical Teeth 48
4.1.2 End Play of Unloaded Double-Helical Pinions 49
4.1.3 Case Study 50
4.2 Equalizing Tilting-Pad Thrust Bearings 52
4.2.1 Pivotal Stiffness of ETBs 52
4.3 The Gyroscopic Reaction Torque 54
4.3.1 High-Momentum Rotors in Imposed Precession 54
4.3.2 Worked Example 55
4.3.3 Rate of Changing Attitude of a Spin Axis 56
4.4 Wobble Plate Mechanics 57
4.4.1 Reaction Torque of a Wobble Plate 58
4.5 The Isoinertial Solid Cylindrical Rotor 59
References 62

5 Interface Sliding, Loading, and Wear 63
5.1 Interface Sliding Per Cycle 63
5.1.1 Pistons of Minimum Skirt Area 64
5.1.2 Engine Valves 65
5.1.3 Couplings for Parallel Offset 66
5.1.4 Hydraulic Distributors 67
5.2 Rolling Versus Sliding Interfaces 69
5.3 Self-Reinforced Traction-Drive Contacts 70
5.4 Homogeneous-Wear Interface Design 71
5.4.1 Continually Varying Exposure of Interface Points 72
5.4.2 Hunting-Tooth Gear Pairs 73
5.4.3 Helical Gears 74
5.4.4 Ball Screw–Nut Systems 76
5.4.5 Numerical Example 78
References 80

6 Elastic Deformation and Microslip Issues 81
6.1 Principle of Matched Elastic Deformations 81
6.1.1 Cascades 83
6.1.2 Keeping Orientation 84
6.1.3 Apportioned Loading 85
6.1.4 Full-Load Bearing 86
6.2 Jointing Detachable Static Interfaces 86
6.3 Fretting Damage in Machines 87
6.3.1 Microslip in Loaded Conformal Contacts 88
6.3.2 Overconstraint-Relieving Joints Prone to Fretting 89
6.3.3 Designing Against Fretting Fatigue 90
6.3.4 Example of Designing Against Fretting Fatigue 91
6.4 Case Study 92
6.4.1 Shaft Failure Characterization 92
6.4.2 Root Cause Analysis 94
References 95

7 Assembly Over Mobile Interfaces 97
7.1 Assembly 97
7.2 Mobility of Mechanical Assemblies 98
7.2.1 Exactly Constrained Spatial Linkages 100
7.3 Kinematic Overconstraints 101
7.3.1 Plurality Overconstraints 103
7.3.2 Angular Overconstraints 103
7.3.3 Lateral Overconstraints 104
7.3.4 Angular and Lateral Overconstraint Combinations 105
7.3.5 Phasing Overconstraints 106
7.3.6 Intrinsic Overconstraints 107
7.3.7 Drawbacks Suffered by Overconstrained Designs 107
7.4 Identifying the Anticipated Mobility 109
7.5 Limitations of Mobility Analysis 110
7.6 Quasi-Exact-Constraint Design 111
7.6.1 Constraint Analysis of a Radial-Piston Motor 112
7.7 Pitfalls in Designing Locked-Train Gearing 113
7.7.1 Interlocked Gear Train with Equal Load Sharing 115
7.8 Spur Gears Made Tolerant of Misalignment 116
References 118

8 Allocation of Functions 119
8.1 Allocation of Functions to Their Carriers 119
8.2 Redundancy 120
8.2.1 Isolating or Decoupling Failed Components 120
8.2.2 Examples of Isolating Failed Components 121
8.2.3 Case Study 122

8.3 Separate Allocation of Functions 124
8.3.1 Examples of the Separate Allocation of Functions 125
8.4 Association of Functions 128
8.4.1 Improper Association of Functions 1 128
8.4.2 Improper Association of Functions 2 130
8.5 The Pitfall of Conditional Functionality 131
References 131

9 Gearing Design for High Power Density 133
9.1 Performance Density Criteria 133
9.2 Split Power Paths 135
9.2.1 Double-Mesh Pinions, Disposition of the Backlash 136
9.2.2 Asymmetric-Power-Split Spatial Gearing 137
9.3 Asymmetric Gear Teeth 139
9.3.1 Planetary Gearing with Asymmetric Teeth 140
9.4 Compound Planetary Gearing 141
9.4.1 Compound Planetary Gearing with Asymmetric Teeth 142
9.4.2 Pitfalls in Designing Asymmetric Planetary Gearing 144
9.4.3 Hydraulic Load Equalizers in Split-Torque Gearing 144
References 146

10 High Power/Torque-Density Devices 147
10.1 Long Shafting of High Power Density 147
10.2 Differential Gearing of Higher Torque Density 148
10.2.1 Assembly of Full-Complement Planet Pairs 149
10.2.2 Worked Examples 150
10.3 Planetary Roller Screws 151
10.3.1 Threaded-Roller Screws 152
10.3.2 Grooved-Roller Screws 154
10.4 Rzeppa Joints of Higher Torque Density 157
10.5 Multi-Disc Clutches 158
10.6 Gear Pumps of High Performance Density 159
References 160

11 Hydraulic Power Density 163
11.1 Routing and Conditioning Mechanical Power 163
11.1.1 Controllability and Efficiency 164
11.1.2 High Power Density Pumps and Motors 165
11.2 Axial-Piston Pumps/Motors 166
11.2.1 Increasing the Power Density of Swashplate Units 166
11.2.2 Tapered-Piston Bent-Axis Units 169

11.2.3 Comparing Same-Displacement Axial-Piston Units 172
11.2.4 Design Review of Pump Bearings 175
11.3 Performance of Hydraulic Pumps/Motors 178
11.3.1 Energy Conversion 180
11.4 Higher-Efficiency Variable Pumps/Motors 181
11.4.1 Off-Center-Pivoted Designs 181
References 184

12 Megawatt-Scale Fluid Power 185
12.1 Transforming Mechanical Shaft Power 185
12.2 Multi-Pump Systems 186
12.2.1 Worked Example 187
12.3 Wind Turbine Fluid-Power Transmission Layouts 189
12.3.1 All the System Components in the Nacelle 189
12.3.2 Hydraulics in the Nacelle, Generator at Ground Level 190
12.3.3 Pump(s) Only in the Nacelle 191
12.3.4 Comparative Assessment 192
12.4 Ripple in Piston Pumps and Motors 193
12.4.1 Optimum Number of Pistons 196
12.5 Mitigating Ripple by Angle Phasing Two Units 197
12.5.1 Phasing Back-To-Back Units 198
References 199

13 Energy Retrieval, Storage, and Release 201
13.1 Kinetic Energy Retrieval Systems 201
13.2 Sequence in Energy Retrieval and Release 202
13.3 Over-Center Pumps/Motors 204
13.3.1 Kinematics of Synchronization 204
13.3.2 Power Conversion 206
13.3.3 State of the Art of Variable-Displacement Units 206
13.4 Hydraulic Accumulators 207
13.4.1 Bladder Accumulators 209
13.4.2 Worked Example 211
13.5 Energy Storage in Mechanical Springs 212
References 213

14 Multi-Attribute Designs 215
14.1 Well-Proven Radial-Piston Motor 215
14.2 High Power Density Bent-Axis Unit 217
14.3 Turboprop Gearbox for Both Directions of Rotation 218
14.4 Curved-Tooth Gears 220
14.4.1 Involute C-gears of Constant Pressure Angle 222
14.4.2 State of the Art 223

14.4.3 Increased Torque Density by Curved-Tooth Gears 223
14.4.4 Gears of Base-Surface Circular Tooth Trace 224
14.4.5 Gears of Pitch-Surface Circular Tooth Trace 226
References 228

Nomenclature

a	acceleration; arm length; ellipse major axis
A	addendum-to-module ratio; area
b	backlash; ellipse minor axis
c	center distance; constant (general)
C	basic dynamic load rating of bearing
d	diameter; bore
D	diameter
e	eccentricity; end play
E	elastic modulus; energy
f	face width; frequency; load/unit length
F	failure probability; force
g	gravitational acceleration
G	shear modulus
i	speed reduction ratio
J	mass moment of inertia
k	factor; fraction (general)
K	constant; factor (general)
L	lead; length
L_{10}	life expectancy with 10% failure probability
m	mass; module
M	moment
n	rotational speed
N	number of pistons, teeth, thread starts
p	pitch; pressure
P	power; equivalent dynamic bearing load
q	number of planetary elements
Q	fluid flow rate
r	radius; pitch-circle radius
R	radius; reliability; ripple
s	stroke
t	thickness
T	torque
v	velocity
V	volume

x profile shift factor
Y thrust factor of tapered roller bearings
z axial distance
α angular acceleration; ball–race pressureangle; swashplate tilt angle
β cylinder block tilt angle
γ specific heat ratio of nitrogen
δ angular misalignment; elastic deformation
η efficiency
θ general angle designation
κ semi-cone angle; spin axis inclination
λ lead angle
μ friction coefficient
ρ density
σ tensile stress
τ shear stress
φ pressure angle in gears
ψ helix angle; tooth trace inclination
ω angular velocity
Ω angular rate of changing attitude

Subscripts

a addendum; axial
b base
B bearing; cylinder block
BA bent-axis
c centrifugal; contra; counteractive
C brake caliper; planet carrier
d displacement; drive; drive-side
e end; external
f fluid; frictional
F face gear; drive flange
g gyroscopic
G gas; gear
h double-helical
in initial; input; inside
L load
m mechanical
max maximum
min minimum
M motor
n normal; nutation
nom nominal
N nut

out	output
0	no-load; offset; overall
p	pitch-plane; precession
P	pinion; planet; plate; propeller; pump
r	radial
re	recirculated
R	ring gear; roller
s	spin
S	screw; sun gear; system
SW	swashplate
t	tangential
v	vertical; volumetric
w	wobbling
xx	transverse-axis

Abbreviations

AM	Assembly Mobility (count)
CR	Contact Ratio
CVT	Continuously Variable Transmission
DOF	Degrees of Freedom
ETB	Equalizing Tilting-pad Thrust Bearing
FE	Finite Element
FOM	Figure of Merit
G	Grübler's count
HCF	Highest Common Factor
I	Idle/Independent mobilities (count)
KERS	Kinetic Energy Retrieval System
M	Mobility (count)
O	Overconstraints (count)
PF	Pitfall Factor
PKM	Parallel Kinematic Machine
SCR	Semi Contact Ratio

Introductory Definitions

1

1.1 Design for Durability

In the mechanical engineering literature and practice it is usual to find *design for durability* almost uniquely synonymous with *design for fatigue endurance*. This may be due to the fact that the majority of failures of mechanical components result from fatigue; bulk material and surface material fatigue. Design for fatigue endurance is based on either one of the following.

- Certified codes or standards for rating the endurance, which codes have evolved from a century-long practice with the design, manufacture, operation, and failure analysis of machine elements such as gears and rolling-element bearings, yet hardly of other elements.
- Stress/strain analysis of components by computer numerical methods such as the finite element (FE) method, with an aim to identify the fatigue hot spots and iron them out in a sense of designing for minimum material usage. The result is sometimes referred to as FE-safe components. There are currently several software tools for the automation of calculations pertinent to the design for fatigue endurance.

However, design for fatigue endurance per se is to belong to the *downstream activities* along the design process, where the type, intensity, duration, and frequency of the cyclic loading are given input parameters. This activity is preceded by implementing higher-level directives and design principles consolidated under the notion *design for durability*. This relies primarily on experience in mechanical design and aims to apply a diversity of measures to improve durability. In particular, this involves opting for the design configuration that subjects the components to smallest possible loads and their mobile interfaces to minimum possible sliding rates (area/volume) for given functional requirements and power rating. Disproportionate loading increases the wear rate of mobile interfaces, initiates surface

H. A. Arafa, *Design for Durability and Performance Density*,
https://doi.org/10.1007/978-3-030-56816-0_1

material fatigue, and hence reduces the equipment durability. It also increases the frictional power loss; lowering the energy efficiency.

1.2 Loads and Loading in Mobile Assemblies

1.2.1 Determinate Loads

Determinate loading in a mechanical system—in terms of force or torque—could have its origin in either one or a combination of some of the following effects.

1. Gravitational, such as in hoisting equipment.
2. Inertial, due to changes imposed on the motion of rigid bodies; applying Newton's second law of motion in any of its forms.
3. Fluid dynamic; hydrodynamic, aerodynamic, or gas dynamic
4. Fluid pressure; hydrostatic or gas pressure, of external supply or internal creation.
5. Magnetic, through electric machines.

For a mechanical system composed of one or more modules to perform its function, a load or more than one identical part loads will be applied in/on it from any of the above origins, which will then be routed through its parts and mobile interfaces to ultimately be grounded or consumed. Grounding, in stationary mechanical equipment, will be in form of mounting reaction on a casing or containment, whereas consumption, in mobile equipment, will be in form of a propulsion effort, on terrain or in air or water.

Loads routed through a mechanical assembly will branch into components of active as well as reactive nature. An active load (force or torque) applied at a point in a mobile assembly is one which, when multiplied into the component velocity at this point in the same loading direction, produces (active) power. But when the component velocity at that point is in the opposite direction to the load, then the result will be negative power; the load being pushed against its *will*.

Some mechanical systems do superpose on the active load a couple of equal and mutually opposite loads of a magnitude depending on their intrinsic amplification property. These will be named *mutually counteractive loads*. They proportionally relate to the active load through static force analysis, and produce so-called virtual or recirculated power in the system; one component of the couple being added to the active load and the other acting oppositely. These loads are considered subtle or obscure loads that could easily escape the designer's attention.

An active load could also be associated with one or more determinate reactive loads; of an inactive nature, which are related to the active load through some trigonometric relations involving the pressure angle (complement to the transmission angle).

Design for durability implies among other things identifying the extent of reactive loads and/or the existence of mutually counteractive loads; such loads that do not necessarily take part in fulfilling the functional requirement of the device to be designed. Design actions should then be taken in order for theses loads to be avoided, balanced out, minimized, or rendered acting on-demand, as appropriate. If not possible, then these loads should fully be taken into account. Negligence of properly dealing with unnecessarily high loads is considered a design pitfall, rather than mere miscalculation.

1.2.2 Indeterminate Loads

An *overconstrained* mobile assembly conceals statically locked-in loads of indeterminate values, even at standstill. These values depend on the accuracy and rigidity of the components. In planar mobile assemblies the locked-in loads are coplanar, but in spatial ones they are, in general, of a twisting and tilting nature; leading to uneven stress distribution in other-than-spherical interfaces and to edge loading in noncircular line contacts. Avoiding or getting rid of the indeterminate loads helps providing autonomous load sharing equalization in multiple load paths and subjecting the components and their mobile interfaces to as even a load/stress distribution as possible.

1.3 Design Pitfall Consequences; Weird Encounters

Design pitfalls are defined by Arafa (2006) as those innocent mistakes that could be attributed to the negligence or ignorance of particular obscure details and/or characteristics of the system and, in some cases, the outcome of manufacturing processes. Designs that conceal such pitfalls are prone to cause weird encounters during assembly or operation of the system. Some pitfalls are of immediate, embarrassing consequences such as failure to assemble, while others lead to short- or medium-term, unexpected failure. Following is an enumeration of the latter category of design pitfalls and their consequences.

1. Presence of subtly amplified loads, an example of which is in power-recirculating planetary gearing systems, which will exhibit poor energy efficiency and be of inferior durability.
2. Presence of parasitic loads, examples of which are the gyroscopic effect in high-angular-momentum spinning parts, and the rotating bending moment due to equalizing tilting-pad thrust bearings when misaligned, ignorance of which means subjecting the system to larger loads than anticipated.
3. Anomalous asymmetry, examples of which are two-stage gearboxes when driven in the *wrong* direction, ball screws with an improper number of track

half-turns, and double-helical gearing with unequal backlashes in the two halves, leading to faster than expected failure.
4. Kinematic overconstrainedness, which results in locked-in stresses and loads of indeterminate values, which are superimposed on the calculated ones and increase the interface stress and wear rate, or in uneven torque distribution in gearing of multiple power paths.
5. Elastic deformation mismatch, which leads to fretting and/or fatigue failure in highly stressed components.
6. Timing or phasing pitfalls, such as if designing hydraulic pumps or motors with an even number of pistons, causing high ripple that in cases would lead to fatigue failure.
7. Improper association or combination of functions, sometimes resulting in conditional functionality of the system.
8. Dependence on uncertain characteristics or variables, with no redundancy, typically causing hard-over failure.

Short-term consequences are encountered when testing/investigating the first prototype of a given design, or else during the warranty period, which will then be costly. Taking action on medium-term consequences (for saving the reputation of the designer and/or the manufacturer) will also be costly; voluntary remedial action such as when recalling cars for fixing some issue.

1.4 Performance Density

The technical expression *performance density* has informally emerged in the early years of this millennium and is being increasingly adopted by the manufacturing industry to advocate some attribute(s) of a product, perhaps in lieu of the less specific adjective *high-performance*. There exists no globally agreed definition of performance density, but it seems to relate—rather qualitatively—a main attribute or specification figure of a product to one of the following dimensional properties.

1. Area, such as footprint.
2. Volume, alternatively expressed as size, installation space (requirement), design space, or design envelop.

Although specification figures are well defined in magnitude and units of measure, it is the qualitative nature of the *attributes*, as well as the non-rigorous expression of dimensional properties (in particular the volume) that will keep the performance density qualitative. However, quantitative expression of the performance density is more nearly possible when a specification figure is related to mass properties of the product. Well-known examples are power density, torque density, and torque-to-inertia ratio. These criteria are useful for a comparative assessment of various designs, makes, or sizes of a given type of equipment; indicating the lightness of a unit relative to its delivering the required performance. It will be noted that sometimes performance density is meant to express the power density, rather; this being a result of translating the German word *Leistungsdichte*, which means both performance and power density.

1.5 Design for Performance Density

According to the above explanation and attempted definition of performance density, adopting the principles of design for durability should be regarded as a key prerequisite for achieving high values thereof. Design for achieving high performance density could not be started before having first considered all the relevant design principles and particulars necessary to achieve high durability, since this would allow the higher stressing, speed, and rating required for the design to qualify for this attribute. Therefore, higher performance density of mechanical systems could be achieved through abiding by the following design directives and principles, as deemed applicable to the specific cases.

1. Implementing or conforming with the principles of design for durability, in general.
2. Opting for dynamically balanced designs in both the rotating and reciprocating machinery to overcome or push forward the limitations on the operating speed due to unduly high loading and vibration; inherent or direct dynamic balancing being preferred to balancing over bearings.
3. Favorably applying the self-aligning principles, particularly in heavily loaded machinery.
4. Design optimization with the aim to maximize such attributes as load-carrying capacity, useful-to-dissipative effort ratio, acceleration/deceleration capability, among various other attributes.
5. Using high strength-to-weight materials such as aluminum alloys and composite materials, instead of ferrous alloys, whenever deemed appropriate.
6. Using high fatigue strength and high surface fatigue strength materials, particularly for Hertzian contacts such as in gears, rolling-element bearings, and continuously variable transmissions (CVTs).
7. Considering multi-element design in lieu of single-element design, such as multiple V-belt drives, multi-strand roller chain drives, and multi-disc clutches.
8. Preferring gear transmission systems of multiple meshing viz. power split gearing in lieu of single-mesh ones.
9. Applying the design principle of *nearest counteraction*—named by Pahl and Beitz (1984) the principle of "short and direct force transmission path"—for achieving maximum rigidity, least number of loaded mobile interfaces, and minimum material usage.
10. Properly apportioning the distribution of material elements in loaded components, such as using closed hollow sections (hollow shafts and helical-tube

springs), ribbing of casings, and minimizing the web material of large gears and rotors.

11. Maximum utilization of space, such as in full-complement designs and in maximum displacement-volume design versions of hydraulic pumps and motors.
12. Aiming at lowest frictional power losses; preferring rolling-friction interfaces to sliding-friction ones, unless the latter operate under a sustained hydrodynamic fluid film.
13. Maximizing the hydrodynamic efficiency of rotodynamic pumps and turbines (outside the scope of the present text).
14. Maximizing the thermodynamic-cycle efficiency of engines (outside the scope of the present text), in addition to implementing the relevant ones of the above measures. Recent examples of high performance density are found in racing car engines.

References

Arafa HA (2006) Mechanical design pitfalls. Proc I Mech E, Part C, J Mech Eng Sci 220(6): 887–899. https://doi.org/10.1243/09544062JMES185

Pahl G, Beitz W (1984) Engineering design. The Design Council, London and Springer, Berlin

Managing Reactive Loads

2

Reactive loads could be either forces or moments that usually do not take part in the functioning of the mechanical equipment. Reaction forces are present as a burden on the bearings or the mobile interfaces in general. Design measures to be taken for enhancing the durability of the mechanical system by reducing or completely balancing these reactions are presented. A real-life case study is also presented that shows how negligence to do so could lead to unexpectedly premature failure, in expensive pieces of equipment. Another case of interest is that of the durability of car front wheel bearings and how it could be enhanced. Unduly high reaction moments are also objectionable for the elastic deformations they produce, among other effects, and, if not carefully eliminated or minimized, would represent some design pitfall. Application of the design principle of minimum pressure angle will be discussed in the same context of durability enhancement.

2.1 Reducing the Bearing Load by Proper Placement

The *proper* placement of some component(s) within the design happens to have a considerable effect on the reactive loads developed in operation, hence on the durability. This notion will be highlighted by considering two examples. First, gear transmissions often include an idler gear or pinion (idler for short) between a driver and driven gear for the purpose of maintaining the same direction of rotation, extending the center distance, or both. They also often include countershafts (intermediate shafts; layshafts) for higher reduction ratio in case of multi-stage systems. The placement or disposition of the idler or the countershaft has a marked effect on the loading of its bearing(s), which characteristic does not seem to be well appreciated; it will be analyzed with an aim to relieve designers from being tempted into design pitfalls regarding the durability of bearings just by neglecting some simple static force analysis. Second, disc brakes are another noteworthy example of

H. A. Arafa, *Design for Durability and Performance Density*,
https://doi.org/10.1007/978-3-030-56816-0_2

the importance of proper placement of the brake caliper for enhancing the durability of the front wheel bearings in cars.

2.1.1 Proper Location of an Idler Gear

A schematic of a gear set of a reduction ratio of about 3 including an idler gear is shown in Fig. 2.1. The idler is shown disposed at alternative locations ranging between the two mirror-image most compact cases; Fig. 2.1a, d. Maintaining the same dimensions for the three gears (represented by their pitch circles), the equal normal tooth loads F_n are drawn inclined to the common tangents to the pitch circles by the pressure angle φ. The idler angle α is defined as the angle subtended between the two lines of centers between the idler and the driving and the driven gears, measured to the side of bearing reaction on the idler. It follows that the angle between the normal tooth load F_n and the idler angle bisector $\theta = 90° - \varphi - \alpha/2$. The ratio of the resultant idler bearing load to the normal tooth load will thus be $F_B/F_n = 2 \cos \theta$.

The two intermediate cases in Fig. 2.1b, c, where all the loads are parallel and where all the axes are coplanar, respectively, feature the highest idler bearing load and will be ignored for the present purpose. It is seen that the idler bearing load with the idler at the proper maximum offset location; being dragged *into* the mesh zone as in Fig. 2.1d, is much smaller than the value with the idler at the improper location; being pushed *away from* the mesh zone as in Fig. 2.1a, everything else being the same, including the bearing loads of the input and output gears. The ratio

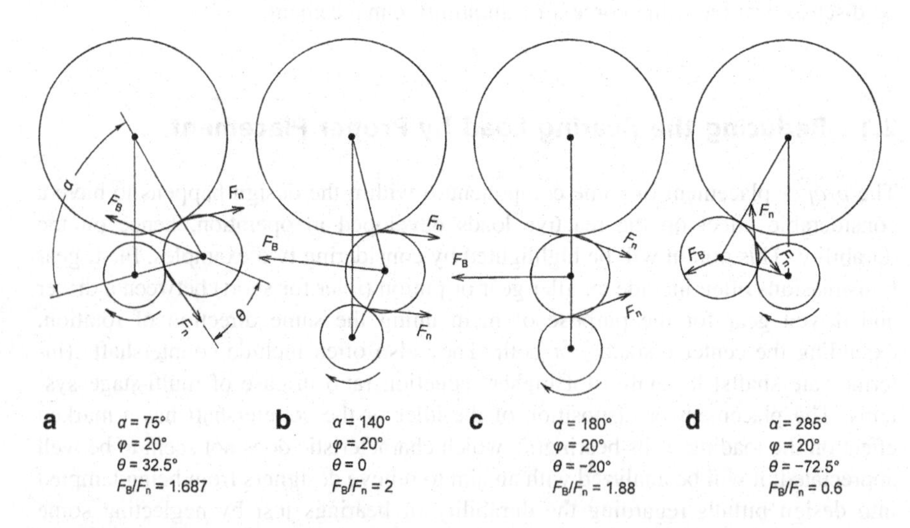

Fig. 2.1 Alternative idler gear locations between a driving pinion and a driven gear, **a** maximum offset in the direction of being pushed away, **b** a little offset that makes the normal tooth loads parallel, **c** coplanar-axis configuration, **d** maximum offset to the opposite side

of idler bearing loads at these two locations (in this example) will be (cos 32.5°)/(cos 72.5°) = 2.805. This ratio would make the life expectancies of the same roller bearings relate by the factor $2.805^{10/3} = 31$ when the idler is placed at the proper and improper locations. It is, therefore, very much worth paying attention to the idler gear, being located at its proper maximum offset location, for unidirectional loading. (The gearing can alternatively be made of direct mesh maintaining the same center distance by making the pinion of one more tooth and the gear of three more teeth; for a reduction ratio of ≈ 3, see Sect. 14.3).

2.1.2 Countershafts in Split-Torque Transmissions

Rotorcraft and turboprop transmissions often include one or more two-stage, split-torque gear sets. These have by necessity two identical countershafts that make the assembly apparently symmetrical, as shown in Fig. 2.2. However, with a relatively small axial extension adopted for compactness, and for a given direction of rotation, the bearings of one of those countershafts (the one that is urged out of mesh) will suffer from the problem of reduced durability, unless that shaft was provided with appropriately larger bearings. The effect becomes more pronounced with closer to unity gear-to-pinion diameter ratio on the countershaft (still at minimum axial separation), where the force analysis approaches that of an idler gear proper. Split-torque gearing is thus, in effect, asymmetric. The drawing features a unity-ratio lever that connects to the non-rotating races of thrust bearings in/on the countershafts to equalize the two axial load components of the helical first stage, hence the power in both branches. This is to avoid the need for fine axial adjustment (and readjustment) of one of the countershafts.

2.1.3 Anomalous Gear Reducers

Two-stage gear reducers are found in numerous industrial applications. Some designs feature collinear (coaxial) input and output shafts, hence referred to as collinear two-stage reducers. They have the following merits.

1. In-line layout of the power transmission.
2. Shaft-mount and flange-mounted versions prevent any angular deflection of the gearbox under load from affecting the alignment of the driveline.
3. Easy setup for machining the bearing bores in the casing (same center distance).
4. Possibility of locating the input-shaft end bearing inside the output shaft.
5. The bearing loading, hence their durability, remains the same in both directions of operation, hence no need for labeling a preferred direction.

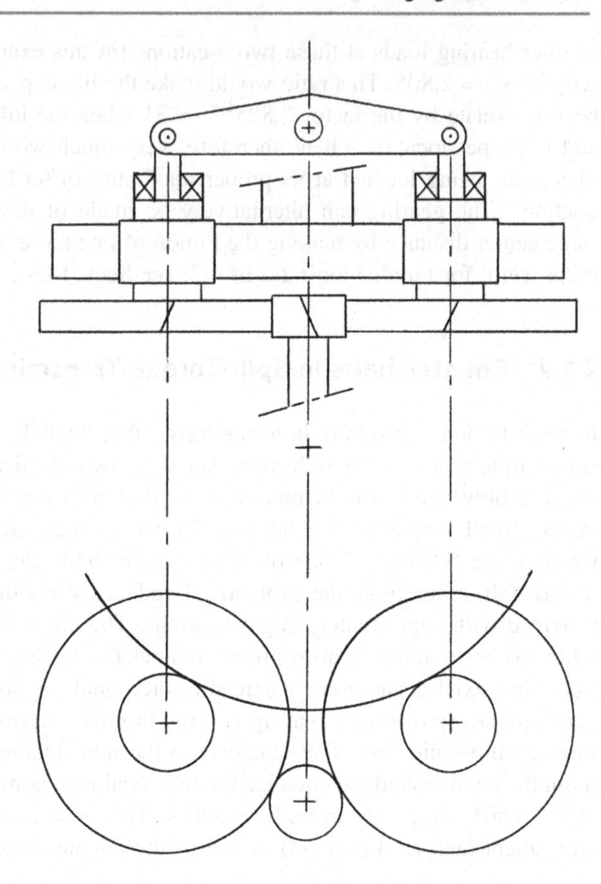

Fig. 2.2 Two-stage split-torque transmission

However, because of the equal center distances of both stages, the choice of the modules and the calculation of the face widths could not be sufficiently compatible with the much different torques in the two stages.

Most appropriate design of both stages could be achieved with a configuration in which the input, intermediate, and output shafts are disposed at the apices of a triangle (sequentially coplanar) for compactness in the transverse plane, such as shown in the schematic in Fig. 2.3 of a shaft-mount reducer; with a hollow output shaft for direct fitting over the driven machine shaft. But this triangular configuration makes *anomalous gear reducers*, with which utmost care should be taken regarding them to be operated only in the favorable drive direction; identified as the one that urges the countershaft into mesh with both the input pinion and the output gear, resulting in a much higher life expectancy of its bearings than otherwise.

However, all the numerous manufacturers of shaft-mount reducers issue power rating tables for their size series for different application factors (service factors), data that is never associated with the favorable/recommended drive direction. This

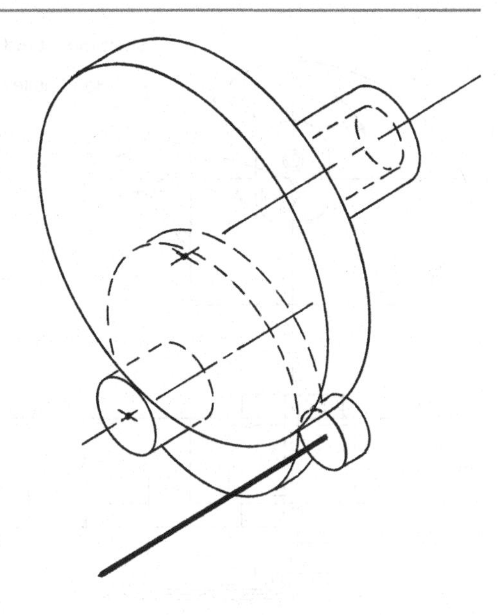

Fig. 2.3 Schematic of a compact two-stage gear reducer

has been going on for decades, and it was and will be left to the application engineer to know or not know of the drawback of anomalous gear reducers; to avoid or to slip into the design pitfall of operating them in the *false* direction.

2.1.4 Case Study

Anomalous shaft-mount gear reducers

Bucket elevators, or belt-and-bucket elevators, are used to lift bulk material such as grain, corn, and the like from ground level up to appreciable heights, to be discharged onto further components of the handling system or into silos. The elevator belt is driven by a head pulley in the head section of the system, and the head pulley shaft is driven by an electric motor, multiple V-belt drive, and gear reducer, at the required speed. A renowned manufacturer of bucket elevators offers his machines in two versions regarding the head pulley drive: right hand and left hand, looking from the back of the discharge spout, as shown in the drawings of the head sections of the units under investigation in Fig. 2.4. These options should offer flexibility in the layout; placing two units side by side, for example. The same two-stage, shaft-mount reducer is used in either case—for the same power rating and discharge height—but the reducer will be operated counter-clockwise in the left-hand version and clockwise in the right-hand version.

Having acquired, installed, and operated a number of those twin side by side bucket elevators, a customer noted that the reducers in the left-hand version failed after some very few months of operation. One of the tapered roller bearings on the countershaft failed early, the other must have followed, both gear meshes became misaligned and severely edge-loaded, and the drive-side gear tooth flanks in both

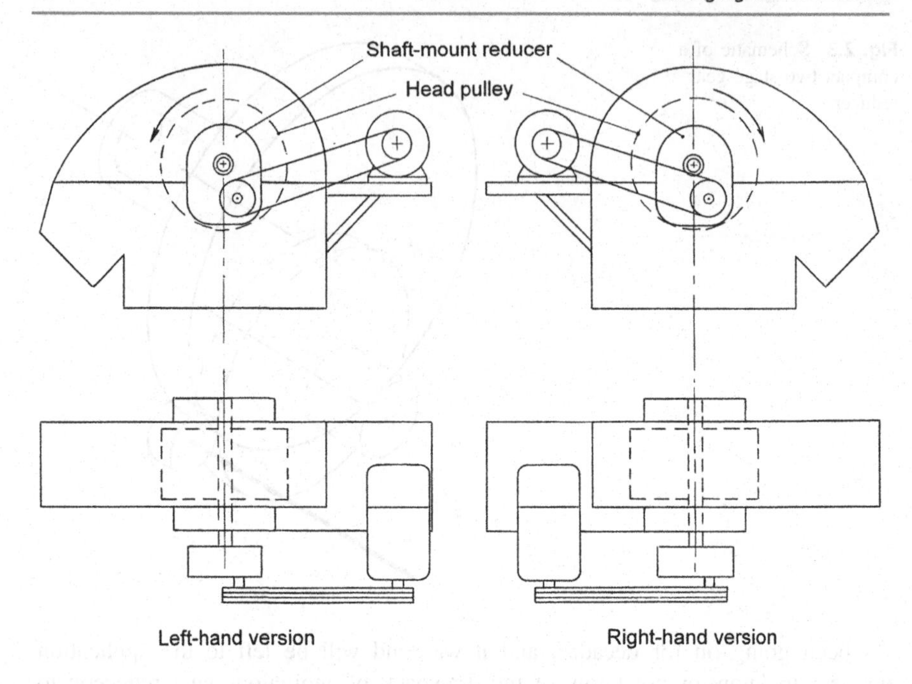

Fig. 2.4 Head section of a bucket elevator in the left-hand and the right-hand drive versions of the head pulley

stages revealed destructive pitting/spalling. Root cause analysis was then made with 3-dimensional force analysis, a summary of which is given hereinafter. The detailed analysis was communicated with the manufacturer who thereupon supplied next-larger reducers, within the warranty period.

Summary of Analysis

Layout drawings of the gearing in the shaft-mount reducer are given to scale—together with the load components on the gears and the bearings—in Fig. 2.5 for the left-hand drive and in Fig. 2.6 in the right-hand drive. Points A and B are the support points of the two identical tapered roller bearings that carry the counter-shaft, which is shown as a double line. The bearings are mounted with their cups face-to-face (X-arrangement).

Essential Data; Specified; Measured

Reducer size (US customary designation): 507 (output shaft diameter 5 7/16 inches).

First stage: center distance = 213.0 mm, teeth ratio 13/66, helix angle = 30.5°
Second stage: center distance = 316.4 mm, teeth ratio 15/75, helix angle = 15.55°
The three shaft centers lie at the apices of an isosceles triangle.

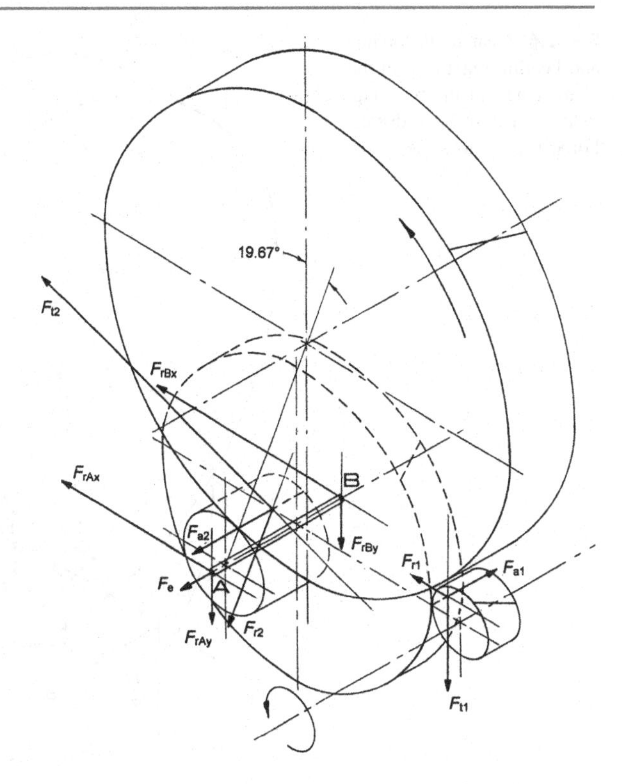

Fig. 2.5 Gear tooth loading and bearing reactions on the countershaft of the two-stage reducer in left-hand drive. Force vectors to scale

Input torque = 570 N.m
Input speed = 1,547 rpm = 162 rad/s (through V-belts from an asynchronous motor); countershaft speed = 305 rpm.
Input power = 92.34 kW ≈ 124 HP (SAE).
Designation of the countershaft tapered roller bearings: K-663/K-653.
Basic dynamic load rating of the countershaft tapered roller bearings $C = 220$ kN (according to the bearing manufacturers' data sheets of that time, around 1990).
Bearing thrust factor $Y = 1.5$

The load components on the countershaft gears and bearings are calculated and plotted to scale in the two cases of counter-clockwise and clockwise input drive. The equivalent dynamic bearing loads P are calculated (taking the circumferential loading conditions of each bearing into consideration), hence the life expectancies

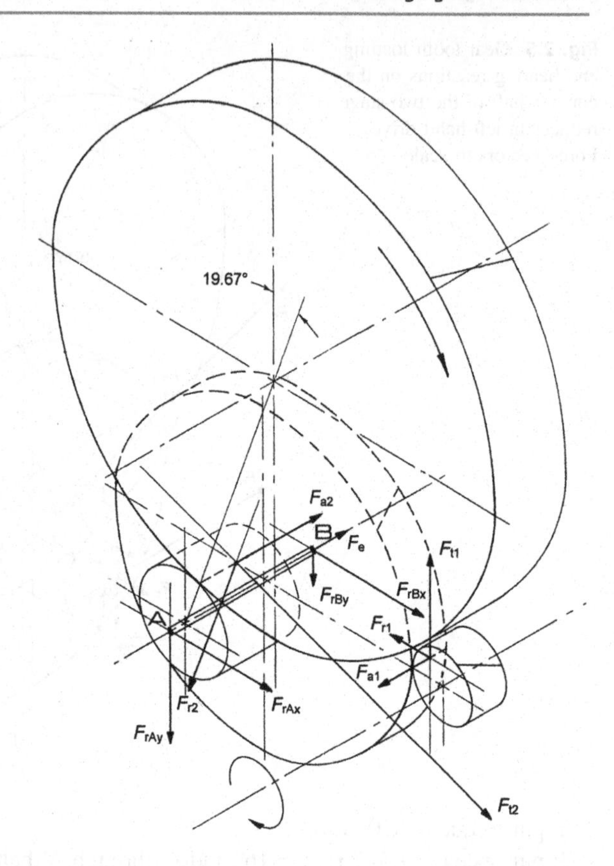

Fig. 2.6 Gear tooth loading and bearing reactions on the countershaft of the two-stage reducer in right-hand drive. Force vectors to scale

(in millions of revolutions) with 90% reliability $L_{10} = (C/P)^{10/3}$. The results are summarized in the following.

Left-hand version; in counter-clockwise drive: P_A = 39,800 N, P_B = 37,000 N.
L_{10A} = 298.6 million revolutions, equivalent to 22.7 months of operation.
L_{10B} = 380.8 million revolutions, equivalent to 28.9 months of operation.

Right-hand version; in clockwise drive: P_A = 24,650 N, P_B = 29,500 N.
L_{10A} = 1475 million revolutions, equivalent to 112 months of operation.
L_{10B} = 810.3 million revolutions, equivalent to 61.5 months of operation.

These results confirm that the left-hand drive reducers (in counter-clockwise drive sense) should be expected to fail that early when transmitting the same power. The results also reveal that even in the favorable driving direction (clockwise) the durability of the two countershaft bearings is not consistent; the gearbox will be expected to endure five years of operation, which limitation is imposed by its bearing B.

Recommendation
Replace with the next larger size (608) reducer, which is the largest available in the series. This necessitates an adaptor sleeve, longitudinally slit to comply with the tapered fixation device in the reducer output shaft, and semi-flat keys since the wall thickness of the sleeve will only be 13.5 mm.

Experience Gained from this Case Study
Careful attention should be paid to two-stage, anomalous gear reducers regarding their proper direction of operation/loading that considerably extends their operational life. The manufacturers are also to be advised to offer their products in right-hand and left-hand versions to suit the different applications, and to advise their customers accordingly.

2.1.5 Disc Brake Caliper Location

Braking puts additional road reaction on the front wheels of a car, so that the front wheel brakes take most of the braking effort. During braking the wheel (together with the brake disc to which it is fastened) is in equilibrium in the vertical direction under three loads, namely the (increased) road reaction F, the vertical component of the wheel bearing load F_{Bv} and the vertical component of the braking force F_C applied by the pads in the caliper, which caliper is rigidly fastened to the wheel spindle. Therefore, the bearing load will be either the sum or the difference of the other two loads, according to the brake caliper being located in front of the axle as shown in Fig. 2.7a or behind it as shown in Fig. 2.7b, respectively. In the first case the caliper force drags down the wheel spindle to increase the bearing load, while in the latter case the caliper force relieves the wheel spindle from part of its load; a substantial amount of the increased car load is transmitted directly to the disc/wheel/road.

The brake calipers of the front wheels should thus better be placed behind the axle to minimize the bearing material distress during hard braking, hence on the long term. This is being implemented in the better engineered, high-end cars; the manufacturers not even having had to answer the question of why. But the majority of car manufacturers do not seem to pay attention to this issue; they just place the brake calipers in front of the axle.

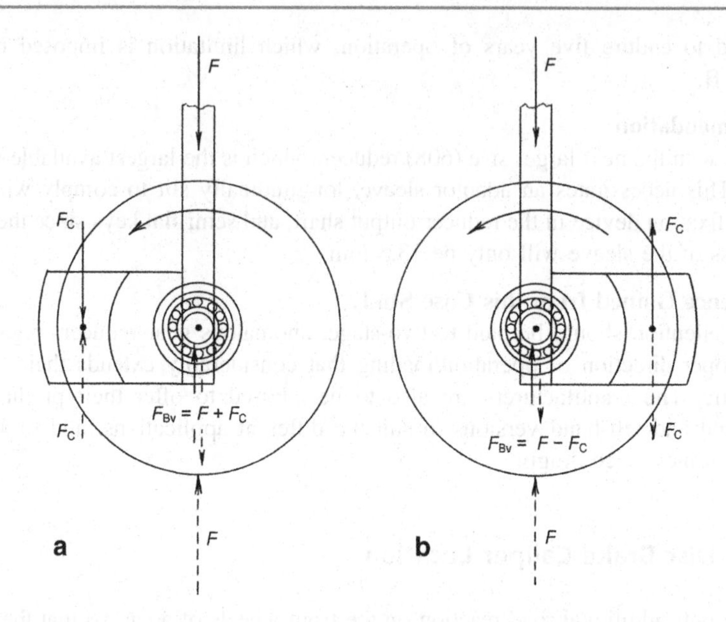

Fig. 2.7 Force analysis during car braking, **a** brake caliper in front of axle, **b** brake caliper behind axle. Solid-line force vectors acting on the suspension strut with spindle, brake caliper, and wheel bearing; dashed-line force vectors acting on the wheel, brake disc, and wheel bearing

2.1.6 Worked Example

Car and Brake Data

Car mass m = 1250 kg.
Equal load distribution on all four wheels is assumed in steady-state driving.
Deceleration in hard braking a = 2.5 m/s^2.
Height of the car center of gravity = 0.3 × wheelbase.
Ratio of mean diameter of brake disc to wheel diameter = 0.5.

Calculation

Steady-state load on a front wheel/bearing = 0.25 mg = 3,065 N.
Hard-braking (horizontal) inertia load = ma = 3,125 N.
Braking inertia load increment/decrement on a front/rear wheel = 0.5 × 0.3 ma = 469 N.
Vertical load on a front wheel in hard braking F = 3065 + 469 = 3,534 N.
Vertical load on a rear wheel in hard braking = 3065 − 469 = 2,596 N.
Horizontal load on a front wheel in hard braking = 3125 × 3534/(4 × 3065) = 900 N.
Tangential braking force on a caliper/disc F_C = 2 × 900 = 1,800 N.

Brake caliper in front of axle: F_C is added to F so that $F_{Bv} = 3534 + 1800 = 5{,}334$ N.

Resultant bearing load $F_B = (5334^2 + 900^2)^{0.5} = 5{,}409$ N.

Ratio to the steady-state load = 5409/3065 = 1.765.

Brake caliper behind axle: F_C is subtracted from F so that $F_{Bv} = 3534 - 1800 = 1{,}734$ N.

Resultant bearing load $F_B = (1734^2 + 900^2)^{0.5} = 1{,}954$ N.

Ratio to the steady-state load = 1954/3065 = 0.6375.

This means that (hard) braking even reduces the front wheel bearing load when the brake caliper is disposed behind the axle. This could simply be interpreted as a considerable part of the front wheel loading being directly routed to the brake caliper, brake disc, wheel, and road, bypassing the wheel bearings.

2.2 Design for Balanced Reactions

Design for balanced reactions, axial and radial, means that any substantial loads should be *dumped* right where they are generated, without being routed (twice) through bearings. Bearings will thus only assume the function of parts location and sustaining the essential loads; those which could not be balanced. Durability of the bearings would thus be much improved, or else their size could be much reduced. Examples of the design for balanced axial reactions are given in the following.

Helical-Gear Drives

The equal and opposite axial reactions of a loaded single-helical gear set could be taken by providing the pinion with a thrust collar of a largely obtuse conical surface, starting outside the addendum cylinder to slide/roll in line contact on a similar feature on the gear wheel side surface below the root cylinder. This well-known idea has first appeared in a patent by Brown et al. (1924). The thrust collar could be provided on one or both sides according to the torque loading being unidirectional or reversed. This relieves the housing bearings of the pinion and the gearwheel of thrust loads; only one bearing will assume the additional function of axial location. Double-helical gearing inherently balances out the axial reactions of the two halves.

Disc Brakes

Disc brakes have an axially floating caliper that contains the hydraulic piston and two friction pads on the two faces of the disc. However large the piston force in hard braking, neither the disc nor the caliper is subjected to axial loads.

Hydraulic Pumps/Motors

Hydraulic axial-piston pumps and motors are sometimes designed in a back-to-back configuration for balancing the heavy axial reactions due to fluid pressure, making the rolling element bearings only support the radial load components on the rotating member. Such designs have been known over the history of fluid power as opposed-swashplate motors, opposed-ball-piston motors, opposed-multi-lobe cam ring motors, and opposed floating-cup pumps and motors (see Sect. 12.5.1).

Turbomachinery and Rotodynamic Pumps
The 1940s witnessed early designs of turbojet engines having the first-stage centrifugal compressor of a back-to-back configuration for balancing the axial air-flow reactions. (The engine should require an elaborate intake duct for routing the air flow in two opposite directions).

Widespread application of the same principle is found in centrifugal pumps for water and other liquids, where a single-sided impeller is subjected to an axial load equal to the mass flow rate of the liquid times its incoming velocity. Starting from units that would be exposed to about 1 kN axial thrust these pumps are invariably designed as symmetrical double-suction pumps to eliminate the need for thrust bearings.

2.2.1 Balancing the Reaction Moments

The transmission of mechanical shaft power from a prime mover to energy consuming equipment—either directly or through a gearbox—is associated with equal and opposite torques on both ends of each transmission shaft. The shaft-end torque of a prime mover or a consumer, and the difference in magnitude of the shaft-end torques on a gearbox produce reaction moments on the respective modules. Heavy reaction moments result in elastic deformations of the casing support and foundation, which moments are usually objectionable, and the connecting shafts would require couplings of extended flexibility to handle the resulting misalignments. Examples of balancing out the heavier reaction moments in power transmission systems include:

1. Dividing the low-speed power into equal halves through two counter-rotating shafts, such as in the drive train of rolling mills, in which the speed reduction and the power-splitting pinions should all be in one casing, rather than using a separate reducer and a pinion stand, as often seen in practice (see Sect. 2.2.2).
2. In cases where only one driving shaft is preferred, a torque tube could be used to connect both casings, surrounding the shaft. However, this tube has to be connected to the casings through the same type of couplings used on the shaft ends to accommodate misalignment, but not to allow rotation.

The transmission of mechanical shaft power from a prime mover to an energy dissipating device, such as an air or water propeller, is associated with a reaction moment on the former. The reaction moment could be nullified by dividing the output power into two equal halves by designing the equipment with.

1. Two counter-rotating shafts, such as in tandem-rotor helicopters.
2. Two contra-rotating shafts, such as in coaxial-rotor helicopters, contra-rotating propeller aircraft, and in some boat and torpedo propellers as well.

Helicopter rotors and aircraft propellers are high-angular-momentum devices; an additional benefit of using contra-rotating shafts in these applications is that the gyroscopic torque is balanced out on the fuselage, which improves the handling quality of these flying machines (see Sect. 4.3.2).

2.2.2 Rolling-Mill Pinion Stands

Rolling mills have customarily been driven through a reduction gearbox and a pinion stand as separate modules in series as shown in Fig. 2.8. The pinion stand contains two or three unity-ratio double-helical gears for the so-called two-high and three-high mills, respectively. The lower shaft is a run-through in the first type and the middle shaft is a run-through in the second type, hence the jargon *straight-through pinion stands*. In either case the pinion stand outputs two equal and opposite torques, so that its mounting reaction moment will equal its input drive torque, just as the torque acting on the output shaft of the preceding reduction gearbox, and no benefit is derived from the counter-rotating output shafts except on the rolling mill itself. Both the gearbox and the pinion stand will require very robust

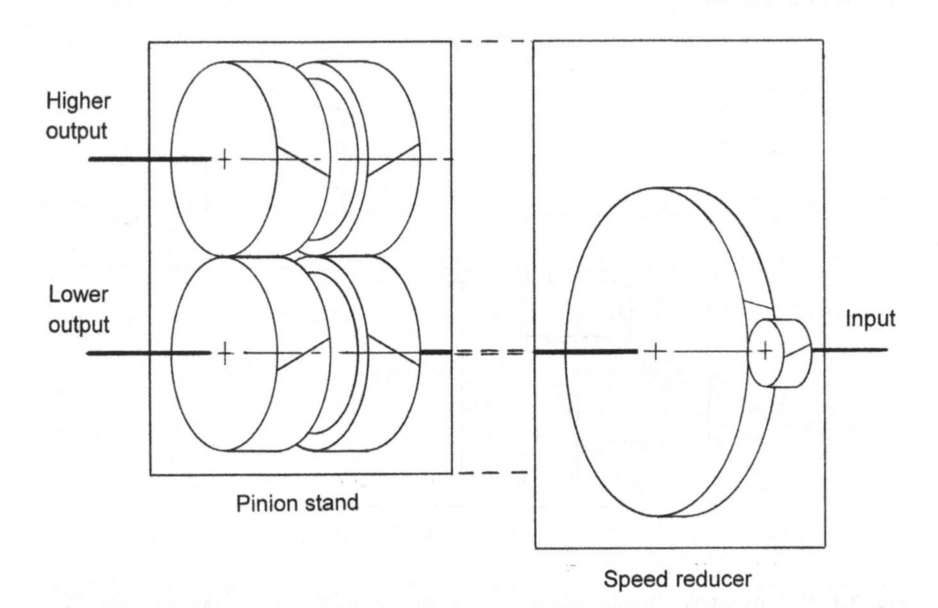

Fig. 2.8 Rolling-mill pinion stand and its preceding speed reducer; in one casing when a rigid connection is established instead of the dashed lines

foundations and holding down means to keep elastic deformations under load to a minimum.

In better designs the speed reduction gearbox and the pinion stand are integrated in one casing as indicated by the dashed lines in Fig. 2.8, thereby reducing its mounting reaction moment to the value of the input drive torque, which is much smaller. The unit is then called integrated or combination pinion stand.

2.2.3 Design Pitfalls with the Reaction Moment

Example 1: Rotorcraft Gearboxes

An input-coupled, high-reduction planetary gearing is shown in Fig. 2.9 after an idea of Platzer (2003) that was intended as a main gearbox for rotorcraft (bearings are not shown). The system features a simple output planetary gear set with the sun and ring gears driven in contra-rotation for achieving the high reduction ratio; an old concept being reinvented, which is a typical power-recirculating system (see Chap. 3). The contra-rotation is realized by a long input spur pinion between two opposed face gears of different size. The large output torque to the helicopter rotor, divided by the mean radius from the rotor axis to the pinion, yields a large lateral force, only to be reacted by radial bearings of the relatively small input pinion, then transferred to the casing (through a flat horizontal cradle, for example) and then to the fuselage. This fact has obviously been ignored; no bearings of this size could take such a reaction.

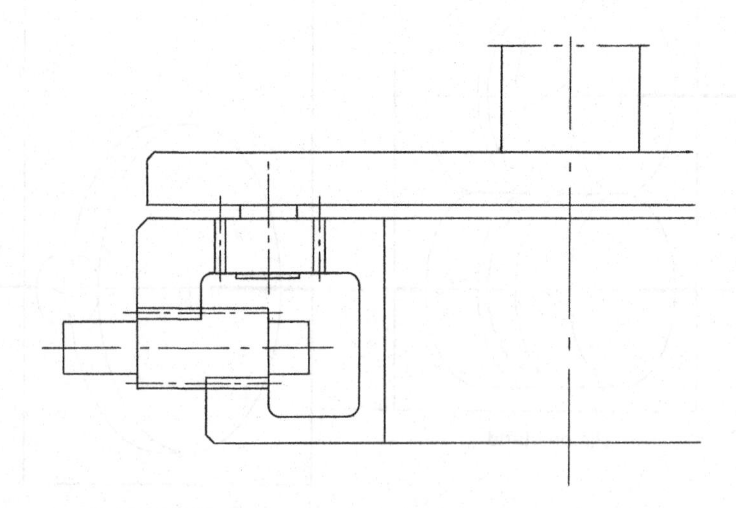

Fig. 2.9 High-reduction planetary gearing with an input coupling stage of two concentric face gears driven in opposition by one spur pinion

Example 2: Machine Tool Main Drives
One suggestion of a headstock design for generating-cutting spiral bevel gears is shown in Fig. 2.10 after a patent by Kreh (2007). The main spindle is driven by a built-in electric motor, while its axis is made to nutate around the vertical axis, tracing out a cone, for the envisaged generating kinematics. This is achieved by supporting the *bent-axis* spindle housing in larger-diameter bearings and rotating it by an electric direct-drive torque motor, of a relatively large diameter as shown. The two motors are therefore *acting in series*; a rare encounter in mechanical design. The fact that the large cutting torque exerted by the main motor is put as a burden on the torque motor to react, in addition to performing its function, must have been overlooked. The torque motor is not intended to fulfill this requirement; it should be a servomotor of the permanent-magnet-rotor type that operates in a precise motion-control mode, and if unduly loaded the gear generating accuracy will be lost.

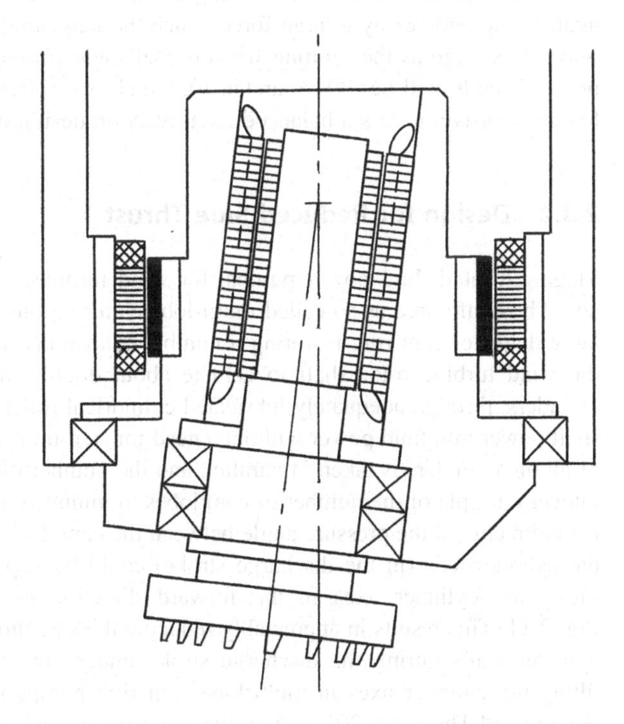

Fig. 2.10 Suggested headstock design of a machine tool for generating-cutting spiral bevel gears

2.3 Principle of Minimum Pressure Angle

Useful forces to be induced between two mating parts in the sense of power transmission are often associated with components acting perpendicular to them. These components are reactive in nature; they produce undesirable effects such as excessive elastic deformations, bearing loads, frictional power losses, and wear. The pressure angle is defined as the angle between the resultant force and the useful one; in the direction required for force transmission. Thus, the smaller the pressure angle the less the undesired reactive force component will be, in principle.

2.3.1 Gearing Versus Traction Drives

The pressure angle is already defined as such in toothed gearing. A typical value is 20°, but 22½° and 25° are also standard values which produce a little larger radial reactive load, but result in better pitting resistance of the gear tooth flanks. On the other hand, a traction drive consisting of two discs requires that they be pressed against one another by a large force, since the tangentially transmitted load could only be as large as the limiting friction coefficient μ times the reactive force. The pressure angle will be $(90° - \arctan \mu)$, which is too large and makes the bearing loads excessive, unless a balanced axial reaction design is adopted (see Sect. 5.3).

2.3.2 Design for Reduced Side Thrust

Megawatt-rated fluid power pumps, for wind turbine transmissions for example, could be configured as so-called multi-lobe cam ring pumps. These are based on a large-diameter cam ring featuring a number of symmetrical lobes, to be driven by the wind turbine rotor shaft to actuate about double that number of pistons in cylinders, through adequately lubricated cylindrical roller followers, to convert the shaft power into fluid power without a need for step-up gearing. The discharge flow of all the cylinders is taken in parallel, and the number of pistons should not be an integer multiple of the number of cam lobes to minimize the ripple. Since the drive is unidirectional the pressure angle between the cam–follower common normal and the cylinder axis (in the discharge stroke) could be kept as small as possible by tilting the cylinder axes in the forward direction as shown schematically in Fig. 2.11. This results in appreciably reducing the side thrust of the pistons on their cylinder walls during the discharge stroke under high pressure. The concept of tilting the cylinder axes in multi-lobe cam ring pumps is originally attributed to Nielsen and Thomsen (2012). A further example of applying the design principle of minimum pressure angle in fluid power pumps and motors is found in Sect. 11.2.2.

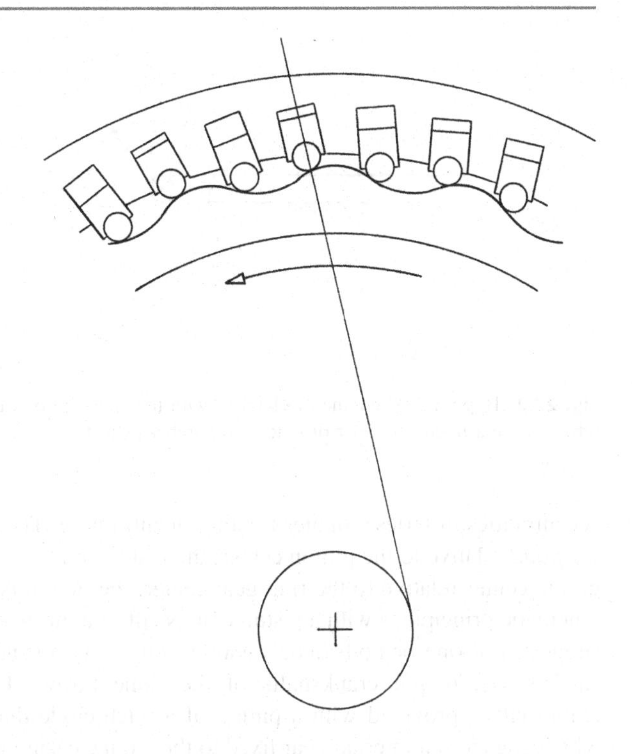

Fig. 2.11 Sector of a multi-lobe cam ring pump with tilted cylinder axes. Fluid ports not shown

2.3.3 Zero Pressure-Angle Engine Mechanism

The well-established piston–connecting rod–crank mechanism of internal combustion engines suffers a little drawback in that the oscillating obliquity of the connecting rod produces side thrust on the piston to bear on the cylinder surface in one direction during the compression stroke and, more heavily, in the opposite direction during the expansion stroke (and to a lesser extent in the other two strokes). This has been seen to cause wear, piston slap, and higher rate of oil pumping into the combustion chamber. The (variable) angle of obliquity or inclination of the connecting rod is the pressure angle between itself and the cylinder axis, which should be kept as small as possible. However, there is an upper limit on the connecting rod-length-to-crank-radius ratio (L/r).

The past 100 years witnessed dozens of inventions and reinventions of engine mechanisms that feature purely translating connecting rods, viz. piston rods, of zero pressure-angle yet still performing the rectilinear-to-rotary-motion conversion. One recent patent on such is by Haynes et al. (2017). These designs are based on the straight-line motion of a point on the pitch circle of a pinion as the latter orbits inside a fixed internal gear of double its size, for the point to describe that special-case hypocycloid, hence the name *hypocycloid engine*, Fig. 2.12. (Any other point on the pinion pitch-circle describes another straight line not aligned with the engine cylinder axis). This mechanism is therefore based on having two equal

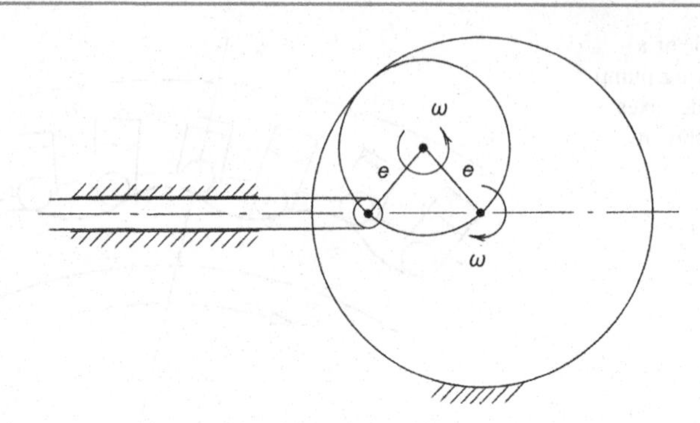

Fig. 2.12 Hypocycloid engine mechanism with the crankpin on the pitch circle of a pinion that orbits inside a fixed ring gear of twice the number of teeth

eccentricities in series, counter-rotating in anti-phase. The one eccentricity is that of the *point* relative to the pinion center; the piston crank, and the second is that of the pinion center relative to the ring gear center; the output crank. Embodiment of this kinematic principle is with a piston crankshaft of a throw e = ¼ the stroke, which is supported at one or both (axial) ends in one or two synchronized pinion (planet) carriers viz. output crankshafts of the same throw. The end(s) of the piston crankshaft is provided with a pinion of a pitch-circle diameter of half the stroke, which engages an internal gear fixed to the engine casing in such an orientation that the pinion eccentricity and the crank-throw vectors add up at top and bottom dead centers. The piston motion will be purely simple harmonic, which could be completely balanced. The major objection to this design, which prevented it from measuring up to a surviving one, is that its main power-handling mechanism is based on gearing, which would be of inferior durability because of the fluctuating torque of each individual cylinder.

References

Brown, Boveri & Cie (1924) Stirnrädergetriebe mit einseitiger Schraubenverzahnung. DE Patent 401652, 6 Sept 1924

Haynes MW, Ramadan ESA, Chassapis C (2017) Planetary crank gear design for internal combustion engines. US Patent 9,540,994, 10 Jan 2017

Kreh W (2007) Machine and method with 7 axes for CNC-controlled machining, particularly generating cutting or grinding, of spiral bevel gears. US Patent 7,179,025, 20 Feb 2007

Nielsen CM, Thomsen JK (2012) Hydraulic transmission. DK Patent 2012 70160, 1 Oct 2012

Platzer M (2003) Planetenkoppelgetriebe für Hubschrauber. DE Patent 102 19 925, 4 Dec 2003

Mutually Counteractive Loads at High Speed

3

The presence of mutually counteractive load torques in the so-called coupling gear set that is included in some configurations of planetary gearing for the purpose of achieving a high transmission ratio represents a problem in gearing design. The thus formed power-recirculating gearing will conceal two sources of relatively high frictional power loss by virtue of the presence of relatively high speeds as well. These sources are the planetary set itself and the coupling gear set, so that the energy efficiency of such gearing could be quite low, and so will be the durability of the gear tooth flanks and the bearings. This is unacceptable when appreciable power is being transmitted. Detailed calculation of the recirculated power ratio, the overload or the so-called pitfall factor, and the sliding conditions hence the frictional power loss in spur gears will be given to come up with a fair estimate of the efficiency of power-recirculating planetary gearing. The systems dealt with in this context are the simple, compound, sunless, and the sunless twinset gearing.

3.1 The Simple Planetary Gear Set

A simple planetary gear set consists of a sun gear, an internal or ring gear, and a number of planets meshing with both, which are supported to rotate on bearings in a planet cage or carrier, usually equally spaced, individually or pair-wise, Fig. 3.1.

Notation N is used for the number of teeth, r for the pitch-circle radius, ω for the angular velocity and n for the speed of rotation in rpm. Suffices are: S for the sun gear, R for the ring gear, and C for the cage or carrier. The number of planets is denoted q.

Assembly Condition To assemble q equally spaced planets the sum $(N_R + N_S)$ should be an integer multiple of q.

Speed Relationships The simple planetary gear set has three terminals (the planet spindles being no terminals) hence it is a two-degree-of-freedom (2-DOF) system. It needs two inputs to produce an output or else, with one input, it produces

H. A. Arafa, *Design for Durability and Performance Density*,
https://doi.org/10.1007/978-3-030-56816-0_3

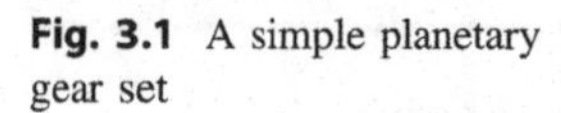

Fig. 3.1 A simple planetary gear set

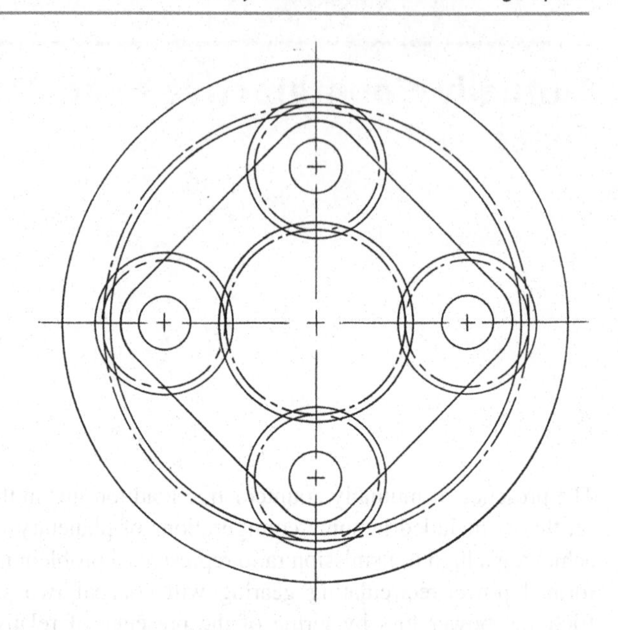

two floating outputs; a differential. Assume all three terminals to rotate at different speeds in one and the same sense; the positive sense. The velocity of the center of the planets is the average of the tangential velocities of the sun and the ring gears, hence

$$\omega_S r_S + \omega_R r_R = 2\omega_C r_C = \omega_C (r_S + r_R)$$

Replacing all ω by n and all r by N to arrive at the general 2-DOF relationship

$$n_C = n_S/(1 + N_R/N_S) + n_R/(1 + N_S/N_R) \quad (3.1)$$

3.1.1 High Transmission-Ratio Planetary Gearing

It has long been recognized that a simple planetary gear set could be designed to give a transmission ratio much in excess of that obtained in its customary configuration (e.g., with a fixed ring gear). This is possible by virtue of the three-term additive expression relating the speeds of the three terminals, Eq. 3.1. High transmission ratio is achieved by adding a gear set that connects with any two terminals of the simple planetary set to couple their speeds—in magnitude and direction—at a little different ratio from the particular one that stalls the third terminal, which is usually the planet carrier or the ring gear. The system will thus be named input-coupled, if it was used as a speed reducer, or output-coupled if it was used as a speed step-up gearing, and the action may be called *differential*.

The three-terminal planetary gear set could be of any configuration such as a planar set (with single or twin planets), a bevel gear set or a face gear set (both with spider carriers), a biplanar set with twin intermeshing planets, or a compound set with two-planet clusters, when the planet of the main set is being made use of as a terminal in addition to the carrier and either the ring or the sun gear. The coupling gearing could also be of any configuration such as a single (offset) double-mesh pinion, a star gear set (with or without reverted gearing), opposed bevel gears, or opposed face gears with an interposed pinion. This gives the variety of coupled-system combinations that have been invented every now and then over the past one hundred years, and which are summarized in the nine configurations in Fig. 3.2.

When loaded, coupled high-transmission-ratio planetary gearing will have certain gear meshes loaded in a driving sense as to transmit power in the throughput direction, and other gear meshes loaded in a driven sense; receiving a smaller power from the system; the recirculated power. The coupling gear or pinion will be the element that handles both powers (at two diametrically opposite mesh points), to equate their difference to the input/output power; to direct the recirculated power back into the system. Figure 3.3 shows the (single) pinion of a coupling gear set, where the driven-sense load and an equal amount of the driving-sense load are said to be mutually counteractive loads F_c (in a torqueing sense), which could be very high, and the (small) difference between them is the net active load F that produces the terminal torque T hence the throughput power $P = \omega rF$. In such gearing configurations a distinct power-recirculation route could generally be readily identified.

3.1.2 Demerits of Power-Recirculating Planetary Gearing

The higher the envisaged transmission ratio the more pronounced the following effects of negative impact on the functioning and performance of the gearing will be.

1. The low-terminal-speed, high-torque planetary gear set will have the planets spinning on their bearings at unduly high speed under the heavy load, so that power loss due to tooth flank sliding friction in both meshes of each of the planets will be high. The conditions in this gear set therefore formally violate the design principle of minimum interface sliding area, particularly under high load, see Sect. 5.1.
2. The high-speed coupling gear set will be loaded (in direct reflection from the planetary set) by much higher than *anticipated*; as when simply calculated from the power and speed data of that coupling set if assumed as a power-splitting or recombining point. This is due to the mutually counteractive load torques; the terminal torque being only the small difference between the two (large) opposed torques, so that it becomes an additional source of a comparably high frictional power loss (The high load on the coupling gear set is sometimes referred to as being result of *intrinsic amplification* of power-recirculating planetary gearing).

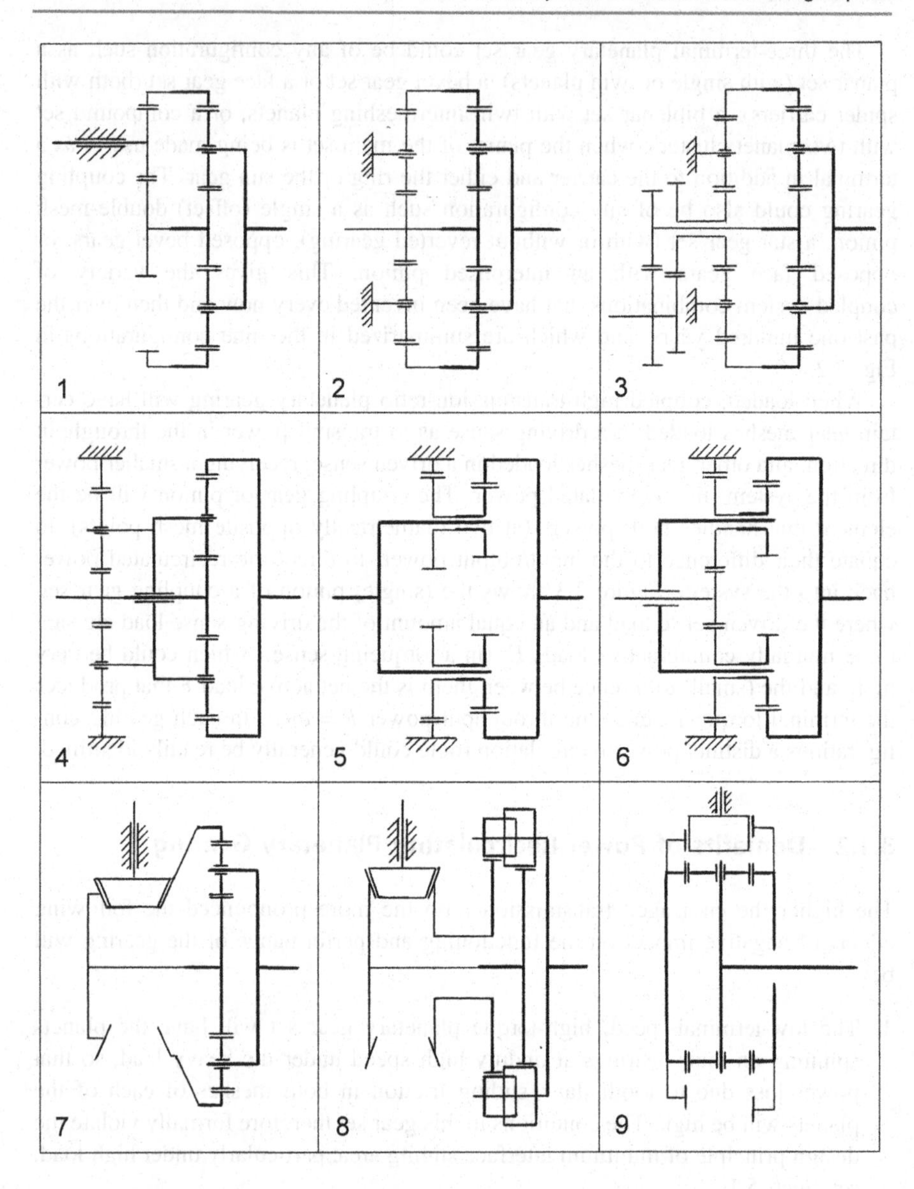

Fig. 3.2 Configurations of coupled gear systems of high transmission ratio. The planetary set with the slow terminal in bold lines, the coupling set at the high-speed end in thin lines

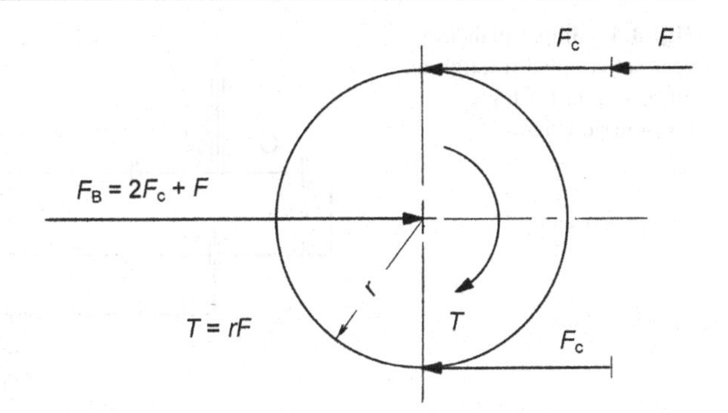

Fig. 3.3 Coupling pinion equilibrium in high-ratio planetary gearing with power recirculation

3. Consequently, the energy efficiency of power-recirculating planetary gearing is too low compared to multi-stage sets. The former should not be used unless a high ratio is needed in transmitting a small power.
4. The actual average tooth loading of the coupling gear set, being much higher than anticipated, should be taken into account when designing that set. The ratio of actual-to-false values of this tooth loading is called the *pitfall factor* (PF).
5. The coupling pinion bearing load, which would wrongly have been assumed negligible, will equal the sum of the two actual tooth loads; $F_B = 2F_c + F$.

As was stated early on by Dudley (1962, p. 3.25), rather qualitatively, "The differential has a large amount of tooth meshing going on at very high tooth loads. Losses are much higher than for regular gearing." With a large ratio of recirculated-to-throughput power the energy efficiency could be very low. However, due to the design compactness and the small number of parts, the concept of coupling the planetary gearing for achieving high transmission ratio has repeatedly attracted design engineers and inventors, possibly without knowing the above disadvantages, just to be tempted into a design pitfall. A good estimate of the power losses is based on knowing the coefficient of friction, the average sliding velocity of the tooth surfaces, and the normal tooth load (see Sec. 3.3.1).

3.2 Coupled Planetary Gearing Characteristics

The basic characteristics of power-recirculating planetary gearing pertinent to loading and power flow will best be highlighted by a numerical generic problem in Sect. 3.2.1. The type of gearing is chosen as the firstly invented reducer by Lissman (1921) which again was reinvented by Wolf et al. (2018), reportedly for application in wind power plants in its step-up mode of operation. This is Type 1 of those coupled high-ratio planetary gear sets given in Fig. 3.2. Another example of power-recirculating planetary gearing is presented in Sect. 3.2.2 and covers further particular issues regarding these systems.

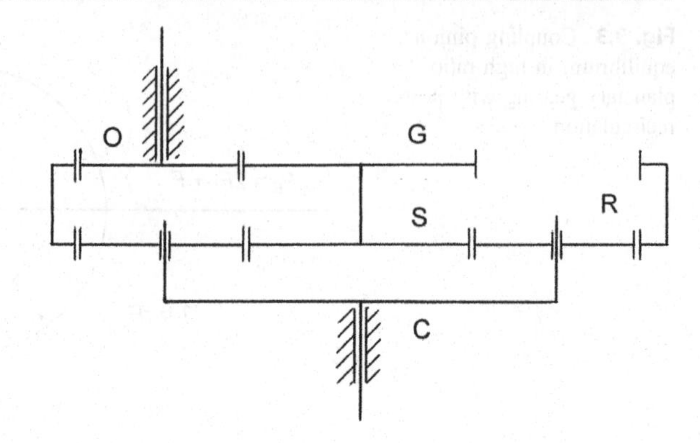

Fig. 3.4 Simple planetary gear set coupled through an offset pinion for high transmission ratio

3.2.1 Generic Problem

In Fig. 3.4 of a simple planetary gear set the rotational speeds of the wide ring gear R and the sun gear S are coupled—by one offset pinion O through an intermediate gear G—to rotate in opposite directions and at such speeds as to give a high transmission ratio between the planet carrier C and the offset pinion; the two terminals of the system. This gearing is thus identified as one that recirculates power.

System Data

Module m = 10 mm for all the gears.
Number of teeth of the ring gear $N_R = 100$.
Number of teeth of the sun gear $N_S = 40$ (thus 4 or 5 planets could be accommodated).
Number of teeth of the offset pinion $N_O = 29$ (for $N_G = 42$).

For a throughput power P = 1 MW at a carrier speed n_C = 29 rpm the following speeds, loads and powers will be calculated in order to identify the extent of the design pitfall with power-recirculating planetary gearing.

Solution

A first iteration is obtained neglecting the power losses; assuming all the gear meshes to be loaded commensurate with the throughput power. Further iterations could follow to refine the results.

Speeds and Loads

$n_C = n_R/(1 + N_S/N_R) + n_S/(1 + N_R/N_S)$.
$n_C/n_O = (29/100)/(1 + 40/100) - (29/42)/(1 + 100/40) = 29/2940$.
Transmission ratio $i = 2940/29 = 101.3793$.
$T_O = P/\omega_O = 10^6/(2\pi \times 2940/60) = 3{,}248$ N.m
$T_C = 3248 \times 2940/29 = 329{,}286$ N.m
Carrier radius $r_C = m(N_R + N_S)/4 = 0.01 \times 140/4 = 0.35$ m.
Sum of planet-bearing loads $= T_C/r_C = 329{,}286/0.35 = 940{,}817$ N.

A planet is a torque-free element, in equilibrium under two equal and parallel tangential tooth loads F_t acting opposite to the bearing load, hence.

$$F_t = 940,817/2 = 470,409\,\text{N}$$

Power Flow

The two tangential tooth loads F_t act in the same spatial direction on the ring and sun gears which have their tangential velocities in opposite directions. The common, floating ring gear is also a torque-free element, hence the offset pinion will be loaded by F_t = 470,409 N in its mesh with it and by $F_{tG} = (40/42)F_t$ = 448,008 N in its mesh with the intermediate gear, both loads being in the same spatial direction, hence in opposite driving directions with respect to the offset pinion.

Offset pinion tangential velocity $v_{t0} = \pi d n_0/60 = \pi \times 0.01 \times 29 \times 2940/60$ = 44.642 m/s

Power flow between offset pinion and ring gear is obtained as

$$F_t v_{t0} = 470,409 \times 44.642/10^6 = 21\,\text{MW}$$

Power flow between offset pinion and intermediate gear is obtained as

$$F_{tG} v_{t0} = 448,008 \times 44.642/10^6 = 20\,\text{MW}$$

which latter amount is the recirculated power (virtual power), to be added to the input power of 1 MW to become the 21 MW flowing from the offset pinion into the system (via the ring gear) in a speed reducing application, or to be subtracted (via gear G) from the 21 MW power for the offset pinion to output 1 MW in a speed step-up mode.

Ratio of recirculated power to throughput power in this system is P_{re}/P = 20.

Pitfall Factor

Should the offset pinion be wrongly considered to drive or be driven by the system from both its mesh points, then the sum of tangential loads on its teeth would have been.

T_0/r_0 = 3248/(0.01 × 29/2) = 22,400 N.

The actual sum of tangential loads on the pinion teeth = 470,409 + 448,008 = 918,417 N.

The pitfall factor is defined as the ratio of the actual to wrongly assumed loads; PF = 918,417/22,400 = 41.

It is thus confirmed that the pitfall factor equals 1 + twice the recirculated power ratio;

$$\text{PF} = 1 + 2P_{re}/P \tag{3.2}$$

It will also be confirmed that the pitfall factor equals the ratio of sum-to-difference of the absolute values of the tangential velocities of the ring and sun gears (only in the present configuration with the planet carrier as one of the system terminals),

$$\mathrm{PF} = (v_{\mathrm{tR}} + v_{\mathrm{tS}})/(v_{\mathrm{tR}} - v_{\mathrm{tS}}) \tag{3.3}$$

Here, PF = (1 + 40/42)/(1 − 40/42) = 41. Equations (3.2) and (3.3) are used for a shortcut solution to problems of power-recirculating planetary gearing, instead of going through the detailed force analysis as in this example.

A further problem with the present system is that the gearbox mounting reaction is taken solely by the bearings of the only one offset pinion; the sum of tangential loads on its teeth, a *heavy* bearing load F_B = 918,417 N as calculated above (When multiplied by the radial distance of the pinion center of 0.01 × (42 + 29)/ 2 = 0.355 m it gives a torque of 326,038 N.m which is the difference between the two terminal-torques 329,286 and 3,248 N.m).

3.2.2 Gearing with Intermeshing Planet Pairs

A planetary gear set with a number of intermeshing planet pairs on the output carrier could be made of a high-reduction ratio when the sun and ring gears are driven in the same sense of rotation at a little different tangential velocity. The ring gear could also be replaced by a second sun gear, in which case the two sun gears should be driven in opposite directions at a little different tangential velocity. The two sun gears could be made identical in size, to be driven at different speeds, or of a little different numbers of teeth, to be driven at equal input speeds. Figure 3.5 is of the latter configuration, where the two sun gears are driven by two identical bevel gears from an input bevel pinion (Type 8 in Fig. 3.2). A so-called *contra planetary reduction ratio* (i_c) could be defined in this particular case as either one of the two equal and opposite input speeds divided by the carrier output speed. It could easily be proved—for any planetary single set operated under such a condition—that

$$i_c = (r_{S1} + r_{S2})/(r_{S1} - r_{S2})$$

This ratio is to be multiplied by the reduction ratio of the input bevel gear stage (in this example) to arrive at the overall reduction ratio of the system.

One pair of intermeshing planets is shown, each on its own pin, in axial view with the speeds, directions of rotation, and the tangential tooth loads (F_t) being indicated (the two planets do not need to be equal). Each planet is a torque-free element, and a meshing pair of such could only be in equilibrium under two forces F_t co-acting on the two planets as shown. The planet carrier is driven by the two (equal) planet-bearing loads, resolved in the same circumferential direction, which simply equal two forces F_t, each being multiplied by its sun gear radius. Hence, the transmitted power is given by

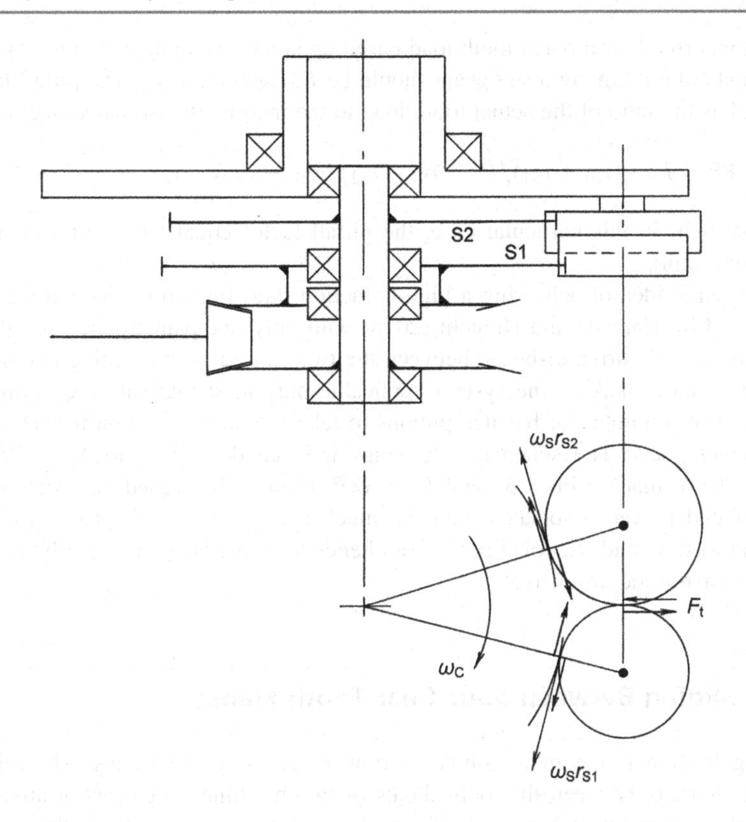

Fig. 3.5 Planetary gear set with two slightly different sun gears and intermeshing planet pairs

$$P = F_t\omega_C(r_{S1} + r_{S2})$$

The force analysis shows that the larger sun gear outputs a power of $F_t\omega_S r_{S1}$ while the smaller sun gear receives back a power of $F_t\omega_S r_{S2}$, indicating substantial power recirculation. The difference gives the throughput power as

$$P = F_t\omega_S(r_{S1} - r_{S2})$$

This is the same as before with $\omega_S/\omega_C = i$. The ratio of recirculated-to-throughput power is then given by

$$P_{re}/P = (F_t\omega_S r_{S2})/[F_t\omega_S(r_{S1} - r_{S2})] = r_{S2}/(r_{S1} - r_{S2})$$

An incorrectly estimated tooth load based upon an assumption of (almost) equal power sharing by the two sun gears should be $P/[\omega_S(r_{S1} + r_{S2})]$. The pitfall factor, defined as the ratio of the actual tooth load to the incorrectly estimated one, is then

$$\text{PF} = F_t\omega_S(r_{S1} + r_{S2})/P = (r_{S1} + r_{S2})/(r_{S1} - r_{S2}) = i_c = 1 + 2P_{re}/P$$

Note that, in this particular case, the pitfall factor equals the contra planetary reduction ratio.

The same idea of achieving a high-reduction ratio for robotic joint drives was suggested by Hamada and Higuchi (2018) with only one planet pair, but with the input bevel gear drive disposed between the two sun gears, preventing continuous output rotation; making the system applicable only to robots, and preventing the insertion of a number of balance pinions to take the heavy reaction torque on the two bevel gears. The schematic drawings indicate data close to $N_{S1} = 95$ and $N_{S2} = 90$, so that i = PF = 37 and $P_{re} = 18P$. It could be argued that such a high recirculated power, associated with the much heavier than anticipated tooth and bearing loading and frictional power loss hence low efficiency, may not be the best choice for robotic arm drives.

3.3 Sliding Between Spur Gear Tooth Flanks

Sliding friction is the main source of power loss in gear systems. The relative sliding velocity between the tooth flanks of two meshing spur gears at any point along the line of action is given by their relative angular velocity times the distance of the contact point from the pitch point, which is the instantaneous center of relative rotation, Fig. 3.6. The sliding velocity is therefore highest at the first point of contact, where it is directed from the tooth tip inwards to the root in the driven gear, and oppositely in the driving gear, as indicated by the arrows beneath the tooth surfaces. As the contact point moves along, the sliding velocity decreases linearly to zero at the pitch point then reverses its direction to increase to another maximum at the last point of contact, where it is directed outwards past the tooth tip of the driving pinion.

For continuous tooth engagement, before one pair of teeth goes out of the mesh zone a succeeding pair should already have come into it. A measure of this overlapping is the contact ratio (abbreviated CR). It indicates the average number of pairs of teeth that share the transmitted load; the extent to which more than one pair exists along the path of contact. The path of contact is the portion of the line of action (tangent to both base circles) which is intercepted by the two addendum circles, Fig. 3.6. Since a length measured along the line of action is equivalent to the same length measured on the base circles, then the contact ratio is defined as the path of contact divided by the base pitch $p_b = \pi m \cos \varphi$, where m is the module. The path of contact is separated by the pitch point into the path of approach, whose length is function of the driven gear parameters, and the path of recess, the length of

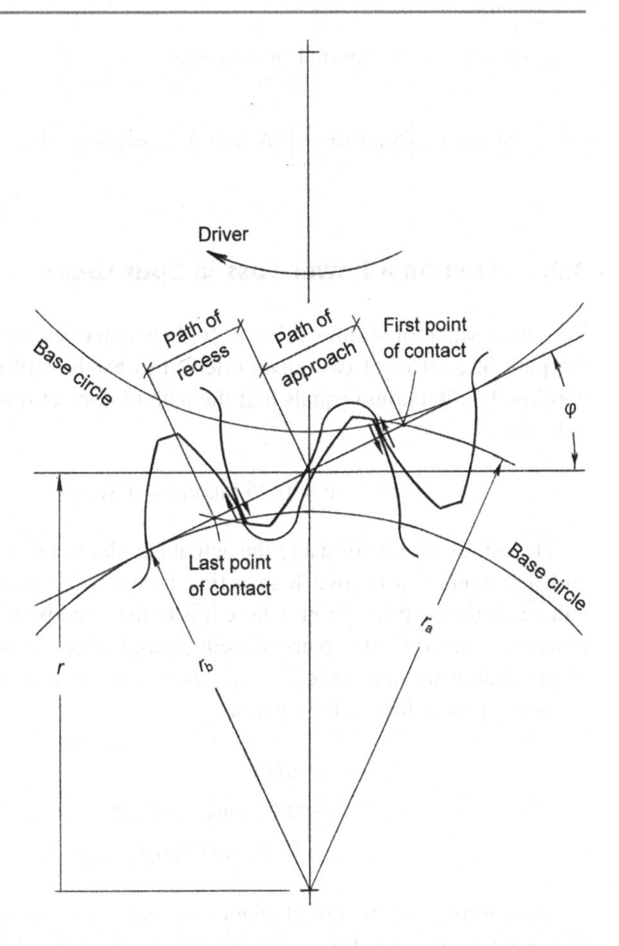

Fig. 3.6 Two spur gears in mesh, showing the relative sliding velocity between the tooth flanks of two consecutive tooth pairs

which is function of the driving gear parameters, only. They switch designation upon switching the function (driving/driven). Thus, it will be more useful to calculate the semi contact ratio (abbreviated SCR) that belongs to each member; the contact ratio being obtained as the sum of the two. For an external gear, pinion, or planet the *path* is given by

$$\left(r_{\mathrm{a}}^{2}-r_{\mathrm{b}}^{2}\right)^{1/2}-r\sin\varphi$$

For an addendum-to-module ratio (*A*) and no profile shift, the semi contact ratio for a gear is given by

$$\mathrm{SCR_G}=\left[\left\{(N\sin\varphi)^{2}+4A(N+A)\right\}^{1/2}-N\sin\varphi\right]\Big/(2\pi\cos\varphi)\qquad(3.4)$$

Similarly, for an internal or ring gear

$$\mathrm{SCR_R} = \left[N\sin\varphi - \left\{(N\sin\varphi)^2 - 4A(N-A)\right\}^{1/2}\right] \Big/ (2\pi\cos\varphi) \tag{3.5}$$

3.3.1 Frictional Power Loss in Spur Gears

The mean value of sliding velocity (v) between one pair of tooth flanks throughout the path of contact of two gears referred to by the suffix numerals 1 and 2 (disregarding the directions) equals half the path of contact times half the relative angular velocity,

$$v = 0.25\,\pi m\cos\varphi(\mathrm{CR})(\omega_1 - \omega_2)$$

The minus sign is for using the actual angular velocity of each gear in magnitude and direction. The normal load acting between the gear teeth F_n assumes its full value around the pitch point, where it is transmitted by only one pair of teeth, and is otherwise shared by two pairs of teeth. It could thus be assumed that the mean value of the sliding friction force is μF_n, where μ is the average friction coefficient. The frictional power loss is then given by

$$\begin{aligned} P_f &= \mu F_n v \\ &= 0.25\,\pi m\cos\varphi(\mathrm{CR})\mu F_n|\omega_1 - \omega_2| \\ &= 0.25\,\pi m(\mathrm{CR})\mu F_t|\omega_1 - \omega_2| \end{aligned} \tag{3.6}$$

Determining the energy efficiency of planetary gearing should be based *only* on this expression. But for a pair of fixed-axis gears it could be divided by the transmitted power $P = F_t\omega_1 mN_1/2 = F_t\omega_2 mN_2/2$ to obtain

$P_f/P = 0.5(\mathrm{CR})\pi\mu(1/N_1 - 1/N_2)$ for internal/external mesh

$P_f/P = 0.5(\mathrm{CR})\pi\mu(1/N_1 + 1/N_2)$ for external/external mesh

These two dimensionless expressions are independent of the load, speed, and module. The energy efficiency is then given by $\eta = 1 - P_f/P$.

Numerical Examples Pair of external spur gears; $\mu = 0.06$, $A = 1$, $\varphi = 20°$, $N_1 = 24$ gives $\mathrm{SCR_{G1}} = 0.8$ and $N_2 = 66$ gives $\mathrm{SCR_{G2}} = 0.9$ from Eq. 3.4, hence CR = 1.7 so that $P_f/P = 0.0091$ and $\eta = 0.9909$. With the smaller gear inside a ring gear with $N_2 = 66$; $\mathrm{SCR_R} = 1.1476$ from Eq. 3.5, hence CR = 1.9476 so that $P_f/P = 0.0049$, and $\eta = 0.9951$. The efficiency of the internal meshing is higher because of the same direction of rotation.

3.3.2 Alternative Estimation of Spur Gear Efficiency

An alternative interpretation of the action of frictional power loss in loaded spur gears could be obtained again by reference to Fig. 3.6. It is seen that, in the instantaneous position when two pairs of teeth share the load; where the distance between them is constantly the base pitch, the sliding friction forces, each of 0.5 μF_n exert a couple of 0.5 $\mu F_n p_b$ that opposes the motion of the driving gear and of the driven gear as well. (The sliding friction forces are indicated by the pair of counterpointing arrows beneath the tooth surfaces). Different conditions prevail during single-tooth contact; around the pitch point. Since relative sliding vanishes at the pitch point, the latter case will be assumed to have a much less effect; to be neglected. This couple is to be added to the driving gear torque and subtracted from the driven gear torque which, when multiplied by their respective angular velocities, will give the net input and output powers. Therefore, the energy efficiency of a pair of external gears will be given by

$$\begin{aligned}\eta &= [(F_n r_{b2} - 0.5\mu F_n p_b)/(F_n r_{b1} + 0.5\mu F_n p_b)](\omega_2/\omega_1)\\ &= (N_2 - \pi\mu)/(N_1 + \pi\mu)(N_1/N_2)\end{aligned} \tag{3.7}$$

For a pair of external/internal gears the friction couple *assists* the motion of the ring gear, such that the minus sign in the numerator (of either expression) becomes a plus sign.

Numerical Examples With the same data of the examples in Sect. 3.3.1 substituted in Eq. 3.7 an efficiency of 0.9894 is obtained for the pair of external gears while (with the plus sign) an efficiency of 0.995 is obtained for a pair of external/internal gears. These results are very close to those obtained previously, which could be regarded as a mutual confirmation of the validity of both schemes.

3.3.3 Worked Example

High-reduction ratios could be obtained from a simple planetary gear set with the sun gear and planet carrier input-coupled, such as to rotate in the same sense; output being from the ring gear. (Zero output speed is obtained when the tangential velocity of the carrier pins is half the tangential velocity of the sun gear). One way of achieving this is to provide an input planetary set with a fixed ring gear, a sun gear on the same input shaft, and a two-sided planet carrier for both sets. This arrangement is shown in Fig. 3.7, being also Type 4 in Fig. 3.2.

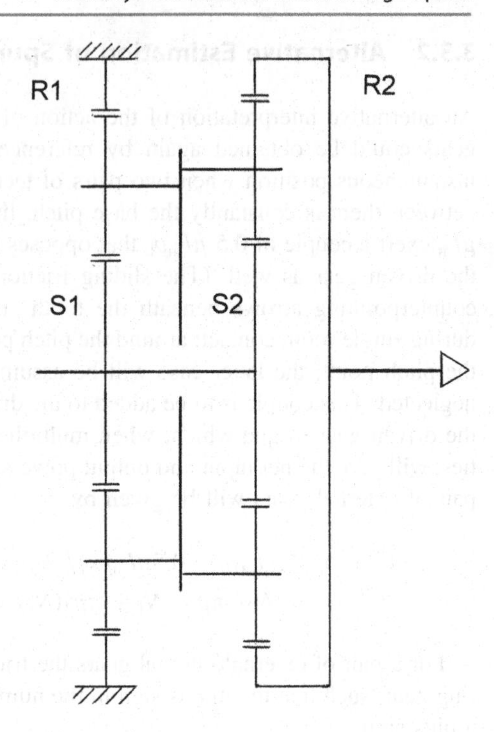

Fig. 3.7 Input-coupled simple planetary gear set of high-reduction ratio with output from the ring gear R2

Data and Requirements

$N_{R1} = 95$, $N_{S1} = 43$, $N_{R2} = 102$, $N_{S2} = 50$ (the planets in both sets are equal and have 26 teeth)
Input speed = 1,425 rpm.
Throughput power = 100 kW.
Module $m = 3$ mm, pressure angle $\varphi = 20°$, and coefficient of sliding friction $\mu = 0.06$.

Assuming, for simplicity, that the loading of all the gear meshes is commensurate with the given throughput power, as if the energy efficiency was 100%, calculate the frictional power losses in each set, and the energy efficiency; as a first iteration.

Solution
The reduction ratio of this system is given by

$$i = (1 + N_{R1}/N_{S1})/(1 - N_{R1}N_{S2}/N_{R2}N_{S1}) = -38.67033$$

First-set contact ratio:

sun/planet = 0.8637 + 0.8104 = 1.6741
planet/ring = 0.8104 + 1.0874 = 1.8978

Second-set contact ratio:

sun/planet = 0.8773 + 0.8104 = 1.6877
planet/ring = 0.8104 + 1.0793 = 1.8897

The relative rotational speeds of the system components are determined by the known tabulation method, and will be omitted for brevity. Only the results are given as follows:

$$n_{\mathrm{P1}} - n_{\mathrm{R1}} = 1178.365\,\mathrm{rpm};\, n_{\mathrm{P1}} - n_{\mathrm{S1}} = 2,603.365\,\mathrm{rpm}$$

$$n_{\mathrm{P2}} - n_{\mathrm{R2}} = 1405.625\ \ \mathrm{rpm};\, n_{\mathrm{P2}} - n_{\mathrm{S2}} = 2,867.475\ \ \mathrm{rpm}$$

Second-Set Power Loss
$F_{t2} = T_2/r_{R2} = 25{,}914/(0.5 \times 0.003 \times 102) = 169{,}372.5$ N (sum of F_t on all the planets).

$$\begin{aligned} P_{f2} &= (0.25\,\pi m\,\mu F_{t2})\left[(\mathrm{CR}\ \Delta\omega)_{S2/P2} + (\mathrm{CR}\ \Delta\omega)_{P2/R2}\right] \\ &= (0.25\pi \times 0.003 \times 0.06 \times 169,372)\times \\ &\quad (1.6877 \times 2867.47 + 1.8897 \times 1405.62) \times \pi/30,000 \\ &= 18.81\,\mathrm{kW} \end{aligned}$$

First-Set Power Loss
$F_{t1} = F_{t2} \times$ carrier ratio = 169,372.5(102 + 50)/(95 + 43) = 186,555 N.

$$\begin{aligned} P_{f1} &= (0.25\pi m\,\mu F_{t1})\left[(\mathrm{CR}\ \Delta\omega)_{S1/P1} + (\mathrm{CR}\ \Delta\omega)_{P1/R1}\right] \\ &= (0.25\pi \times 0.003 \times 0.06 \times 186,555)\times \\ &\quad (1.6741 \times 2603.36 + 1.8978 \times 1178.36) \times \pi/30,000 \\ &= 18.23\,\mathrm{kW} \end{aligned}$$

Total frictional power loss = 18.81 + 18.23 = 37.04 kW
Energy efficiency of the system η = 63%
It should be mentioned that the efficiency of the same power-recirculating planetary gearing configuration with i = 41.1 has been computed by Marsch and Flanagan (2008) using a software package developed for that purpose, giving an unacceptably poor η = 61.6%, thus roughly confirming the above result.

3.4 Compound Planetary Gearing of High Ratio

In addition to the category of high-ratio planetary gear sets that feature a distinct power-recirculating route, there is another category of the same in which no such route could readily be identified, in which the high tooth loads and mesh velocities produce so-called virtual power. This category is the compound planetary gearing that comprises two planetary sets with relatively large, close-to-unity-ratio planets in a cluster; close-ratio planet clusters for short. Figure 3.8a shows the sunless, three-terminal version and Fig. 3.8b the four-terminal version; with one of the sets having a sun gear, where the speeds of the latter and of the carrier being proportioned (the system still being of two DOF). These are Type 5 and Type 6, respectively, in the classification of those coupled high-ratio planetary gear sets in Fig. 3.2. Compounding the two gear sets is achieved by (1) having a common planet carrier, and (2) rigidly attaching the close-ratio planets of the two sets to one another, in gear clusters. The two ring gears will also be in a close-to-unity ratio. The one set with fixed ring gear serves to adjust the relationship between the orbiting and spinning speeds of the planet cluster to a small difference from the condition of stalling the other ring gear. Result is that both planet–ring sets will be loaded to about the same high mutually counteractive tooth loads (the planet cluster being a torque-free element), while all the mesh speeds are just as high as in the low-torque end. This will cause a large virtual power and frictional power loss, and the energy efficiency will be unacceptably low. The two planets in a cluster should also be made closest to one another in axial direction for the cluster not to be under a high tilting moment.

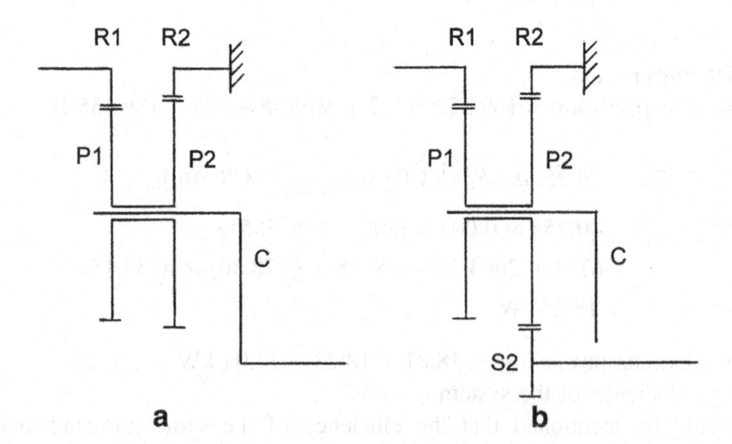

Fig. 3.8 Compound planetary gearing of high transmission ratio, **a** sunless version, **b** with a sun gear

A further drawback of compound planetary gearing—in general—is the necessity of accurately angle phasing every two planets in a cluster such as to carry an equal load share (see also Chap. 7). This adjustment requires well-thought design details and will anyway be an additional burden on the system.

Despite the fact that should be known about the inferior efficiency of compound planetary gearing of high transmission ratios, a suggestion was made by Fernandez Garcia (2009) to use a gear drive of the configurations in Fig. 3.8 for speed step-up applications such as in wind turbines of MW-scale power. Input is at the smaller ring gear (closer to the wind rotor hub) and output is from the planet carrier in the sunless version, or from the sun gear meshing with the larger planets in the second version, in which the carrier is left floating. The latter version is of a much higher step-up ratio. Relatively large planets are essential, allowing only three clusters to be used and leaving a space for a small sun gear. This version is sometimes referred to as Wolfrom gearing.

Previous Experimental Investigation

The energy efficiency of a similar compound planetary gear reducer of a laboratory scale has been experimentally investigated by Bekircan (1987) and found to lie in the range 50–60%. This rather low efficiency was attributed to power "circulation." However, the large power losses should have also been due to the following issues.

1. The results include bearing and oil churning losses as well.
2. The surface finish of the gear tooth flanks may have not be as high as it should be, giving rise to a coefficient of sliding friction larger than anticipated.
3. Inaccurate angle phasing of the planet clusters; it should be known that compound planetary gearing is highly over-constrained, which leads to large locked-in tooth loads and stresses that result in increased frictional power losses.

3.4.1 Sunless Planetary Gearing

A fairly high-reduction ratio is achievable with either gear being grounded in a single set of internal–external gears that have the smallest possible difference between their numbers of teeth, which just allows meshing the planet with the ring gear, in axial direction only, without tip interference. Figure 3.9 shows such an arrangement; the scale drawing ranging from the fully engaged tooth/space along the eccentricity vector to one side up to the teeth on the verge of full clearance, showing how tip interference is just evaded (before the two addendum circles intersect). The condition for avoiding tip interference is detailed in the German standards DIN 3993 Part 1 (1981, Fig. 17) and allows for the ring gear of standard proportions and 56 teeth a planet of a maximum of 48 teeth as shown, with a negligibly small tip rounding. This gives a transmission ratio of 6 or 7, according to which member is grounded, the ring gear or the planet, respectively.

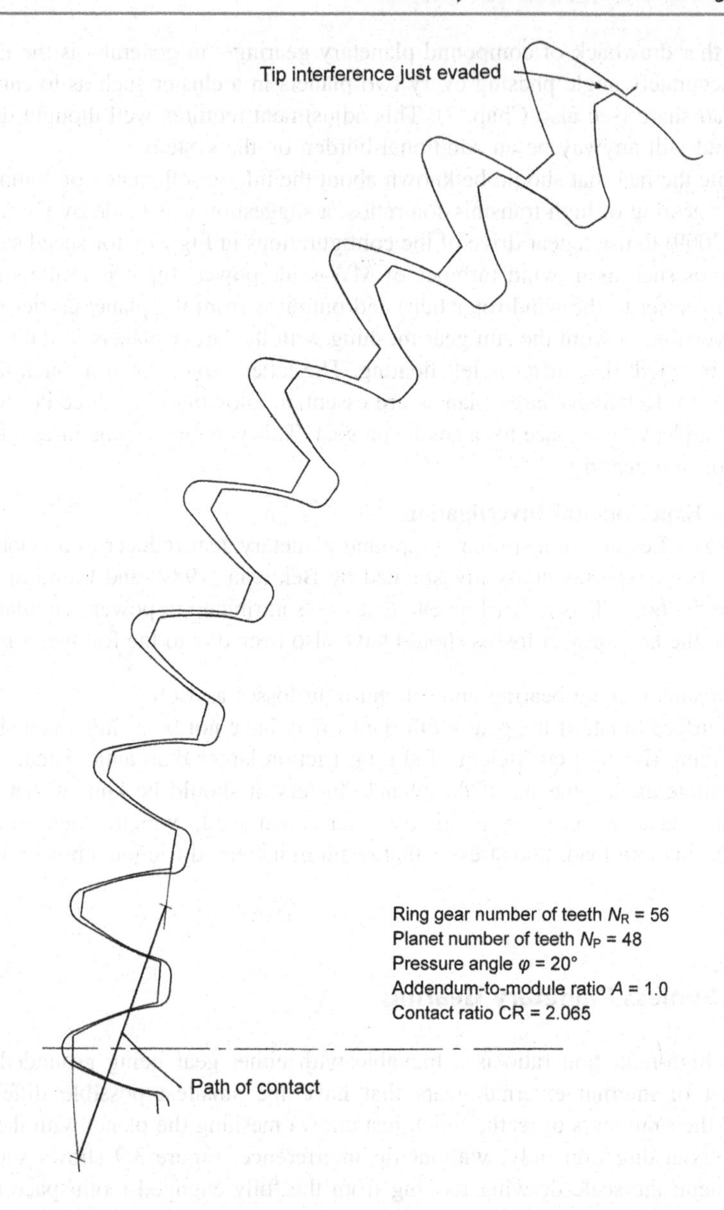

Fig. 3.9 Evading the tip interference between an internal and external gear with the smallest possible difference of their numbers of teeth, without profile shift

But a mechanical filter is needed to connect with the planet, either to extract its slow rotation from the combination of orbiting and rotating motion in case it would be the output terminal, or to prevent its rotation in case the output was from the ring gear. This filtering could be achieved by an Oldham coupling, a quad-link coupling, a double Cardan shaft, or a planocentric drive (see Fig. 7.2b). The latter version of a sunless planetary gearing was proposed in one of the earliest patents by Wildhaber (1930). Furthermore, with a proper choice of the pressure angle and the profile shift(s), internal–external gears with only one tooth difference (in thirty and more) are possible. The reduction ratio will then equal the number of teeth of the non-fixed member. This also makes the eccentricity of the input crank that supports the planet, hence the required balancing mass smallest.

The energy efficiency of sunless planetary gearing is high since the frictional power loss is proportional to the relative angular velocity of the two gears, which is small in ratio to the crank velocity. (This will be evident from the worked example in Sect. 3.4.3). The set could thus be called quiescent-mesh gearing, for the very small tooth flank sliding. The power losses in the bearings and the mechanical filter could even be higher than the tooth-flank frictional losses. However, the torque hence the power density of sunless planetary gearing is limited since a large torque results in a single, large unbalanced rotating-vector tangential tooth load that would be detrimental to the whole assembly, particularly if it was of a substantial size.

3.4.2 Sunless Planetary Twinset Gearing

The attributes of very high-reduction ratio, still with good energy efficiency, and absence of orbiting-motion mechanical filters could all be achieved with sunless

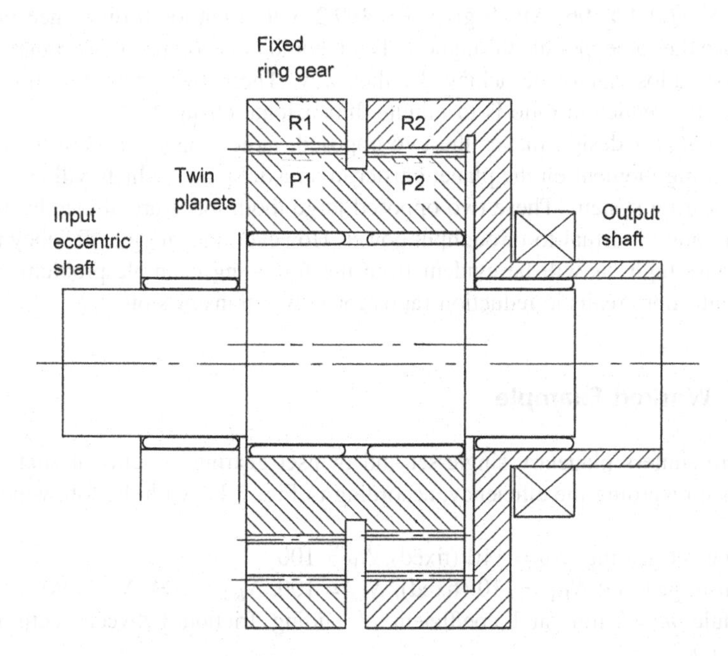

Fig. 3.10 Sunless planetary twinset gearing

planetary twinset gearing. This consists of two single sets of equal eccentricities, compounded by having one common input crank and by attaching the two planets to one another. In an alternative version one planet is an internal gear that will orbit outside a central external gear viz. *sun*.

The system outputs a slow speed from one of the non-orbiting members, with the other held fixed. The basic idea according to the latter version is as old as the patent by Regan (1895). Sunless planetary twinsets have ever since been reinvented dozens of times; the original version with an external-tooth planet cluster as shown in Fig. 3.10 being more common. Denoting the fixed ring gear R1, with output from the second ring gear R2, then the reduction ratio is given by

$$i = 1/[1 - (N_{R1}N_{P2})/(N_{P1}N_{R2})]$$

The reduction ratio could thus be made infinite when the ring-to-planet ratios of the numbers of teeth are equal for both (unequal) planetary sets. Example is a ratio of 9/8 obtained with the combination of numbers of teeth (108 × 88)/(99 × 96). The highest finite reduction ratios are then obtained when each of the four numbers is decreased by one (for positive i) or increased by one (for negative i), keeping the ring-to-planet tooth differences the same. The drawing proportions in Fig. 3.10 could feature a reduced combination of numbers of teeth (107 × 87)/(98 × 95) which gives $i = 9310$ but requires—for equal eccentricities—that the module of the second set be 9/8 times that of the first set, e.g., $m_1 = 2$ mm and $m_2 = 2.25$ mm. Alternatively, just the same drawing proportions could feature the combination (106 × 94)/(104 × 96) which gives $i = 499.2$ with a ten-tooth difference in each set, hence the same module throughout. But it could be recognized that round-figure reduction ratios cannot be achieved either way. These two examples show how sensitive the reduction ratio is to minute dimensional changes.

The *biplanar* design makes the two opposed, large tangential tooth loads produce a tilting moment on the planet cluster in an axial plane, which will be reacted by the bearing system. These two opposed large tooth loads are also indicative of the high ratio of virtual-to-throughput power. However, the energy efficiency is still remarkably high, as will be evident from the following example problem, which deals with more realistic reduction ratios for power transmission.

3.4.3 Worked Example

Compare sunless planetary single-set and twinset gearing systems of small tooth difference regarding the efficiency, according to Fig. 3.11, with the following data:

Single-set gearing: $N_R = 110$ (fixed), $N_P = 100$
Twinset gearing: $N_{R1} = 110$ (fixed), $N_{P1} = 100$, $N_{R2} = 100$, $N_{P2} = 90$
Module $m = 2$ mm and coefficient of sliding friction between teeth flanks $\mu = 0.06$.

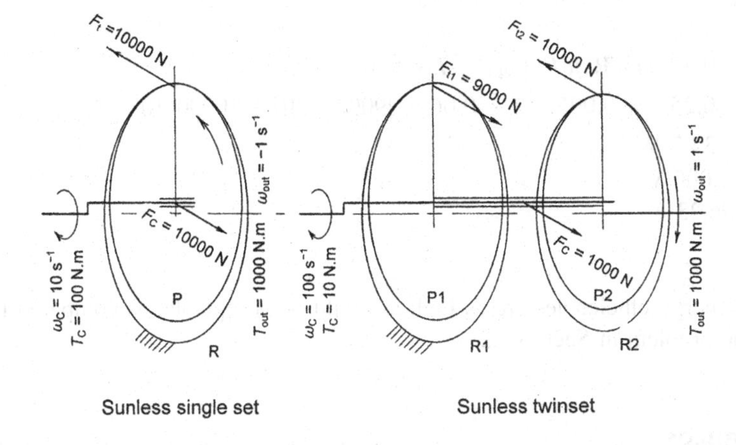

Fig. 3.11 Drawing for the Worked Example 3.4.3

Assume a power of 1 kW to be output at $\omega_{out} = 1\ s^{-1}$ in both cases, and that the loading in the different meshes is commensurate with the throughput power; to obtain a first-iteration value of the energy efficiency.

Solution
Sunless Single-Set

$i = -1/(N_R/N_P - 1) = -10$
$\omega_C = 10\ s^{-1}$, $\omega_{out} = -1\ s^{-1}$
$T_C = 100$ N.m, $T_{out} = 1{,}000$ N.m
$F_t = 10{,}000$ N
$F_C = 10{,}000$ N
CR $\approx 1.07 + 0.93 = 2.0$
$$P_f = 0.25\pi m(\mathrm{CR})\mu F_t|\omega_1 - \omega_2|$$
$$= 0.25\pi \times 0.002 \times 2 \times 0.06 \times 10000 \times 1 = 1.885\ \mathrm{W}$$
$P = 1{,}000$ W.
$\eta = 99.81\%$

Sunless Twinset

$i = 1/(1 - N_{R1}N_{P2}/N_{R2}N_{P1}) = 100$
$\omega_C = 100\ s^{-1}$, hence $\omega_{out} = 1\ s^{-1}$
$T_C = 10$ N.m, $T_{out} = 1{,}000$ N.m
$F_{t2} = 10{,}000$ N.m, $F_{t1} = 9{,}000$ N.m
$F_C = 1{,}000$ N
$\mathrm{CR}_1 \approx \mathrm{CR}_2 = 2.0$

$$
\begin{aligned}
P_{\mathrm{f}} &= 0.25\pi m(\mathrm{CR})\mu[F_{\mathrm{t1}}|\omega_{\mathrm{P1}}| + F_{\mathrm{t2}}|\omega_{\mathrm{P2}} - \omega_{\mathrm{R2}}|] \\
&= 0.25\pi \times 0.002 \times 2 \times 0.06 \times [9000 \times 10 + 10000(10 + 1)] \\
&= 37.7\ \mathrm{W} \\
P &= 1{,}000\ \mathrm{W} \\
\eta &= 96.23\%
\end{aligned}
$$

The energy efficiencies are as high as could be acceptable, as compared to the example problem in Sect. 3.3.2.

References

Bekircan S (1987) Performance of a high-reduction planetary gear drive with respect to efficiency . Proc I Mech E 201 Part D, J Automobile Eng (4):293–297

DIN 3993 Part 1 (1981) Geometrical design of cylindrical internal involute gear pairs; general bases. Beuth, Berlin

Dudley DW (1962) Gear arrangements. In: Dudley DW (ed) Gear handbook. McGraw-Hill, New York, pp 3-1–3-44

Fernandez Garcia A (2009) Power transmission with high gear ratio, intended for a wind turbine. WO Patent 2009/063119, 22 May 2009

Hamada A, Higuchi J (2018) Speed reducer and robot. WO Patent 2018/074008, 26 Apr 2018

Lissman M (1921) Speed reduction gearing. US Patent 1,376,954, 3 May 1921

Marsch J, Flanagan A (2008) Epicyclic gearing: a handbook. Gear Solutions Magazine, Sep 24–31

Regan DS (1895) Power transmitter. US Patent 546,249, 10 Sep 1895

Wildhaber E (1930) Gearing. US Patent 1,767,866, 24 June 1930

Wolf D, Beck S, Kurucz M (2018) Epicyclic gearing with a high transmission ratio. WO Patent 2018/104035, 14 June 2018

4 Subtle and Obscure Loading Sources

The mutually counteractive loads that develop inside power-recirculating planetary gearing as outlined in Chap. 3 represent one important example of the so-called subtle or obscure loads. However, there are other cases to be highlighted. A real-life case study is presented of the premature and unexpected failure due to the designer failing to anticipate the phenomenon of double-helical pinion shuttling, which—as such—is known to gearing design and operation experts. Another source of obscure load is the equalizing tilting-pad thrust bearings when operated under some angular misalignment; they do possess a so-called pivotal stiffness. Gyroscopic design engineers will not be mistaken about the sometimes substantial reactions on the bearings and the casing, but it is feared that such loads may be subtle or obscure to the mechanical design engineer when dealing with fast-spinning, high-angular-momentum machine parts. Essential calculations pertaining to wobbling machine parts are given.

4.1 Double-Helical Pinion Shuttling

Manufacturing a pair of double-helical gears is completed in four separate settings of the cutting and finish-grinding machines; for the four half-gears, with tolerances specified on the outcome of each process. This causes rather wide tolerance limits on the backlash with the gears meshed at correct center distance; the actual backlash being inevitably unequal in their two halves, which condition is referred to as apex mismatch. In one-off or small batch production this anomaly will be more pronounced. As a consequence, reversing the loading direction in operation of a double-helical gear set could trigger *shuttling* of the axially floating pinion.

The severity of the anticipated impulses or *kicks* depends on several factors, such as the values of the two backlashes, the difference between them, the helix angle, and the speed of operation. They are so-called *parasitic, hammering effects* in the system which, when repeatedly applied under certain circumstances, could have

H. A. Arafa, *Design for Durability and Performance Density*,
https://doi.org/10.1007/978-3-030-56816-0_4

loosening effects with consequential damage. Negligence of the possibility of this to happen is a design pitfall.

4.1.1 Relative Positions of Double-Helical Teeth

The successive possible positions of a double-helical pinion tooth relative to the tooth space of the mating gear are shown in Fig. 4.1 with unequal backlashes between the two half gear sets, in operation when the pinion driving power is

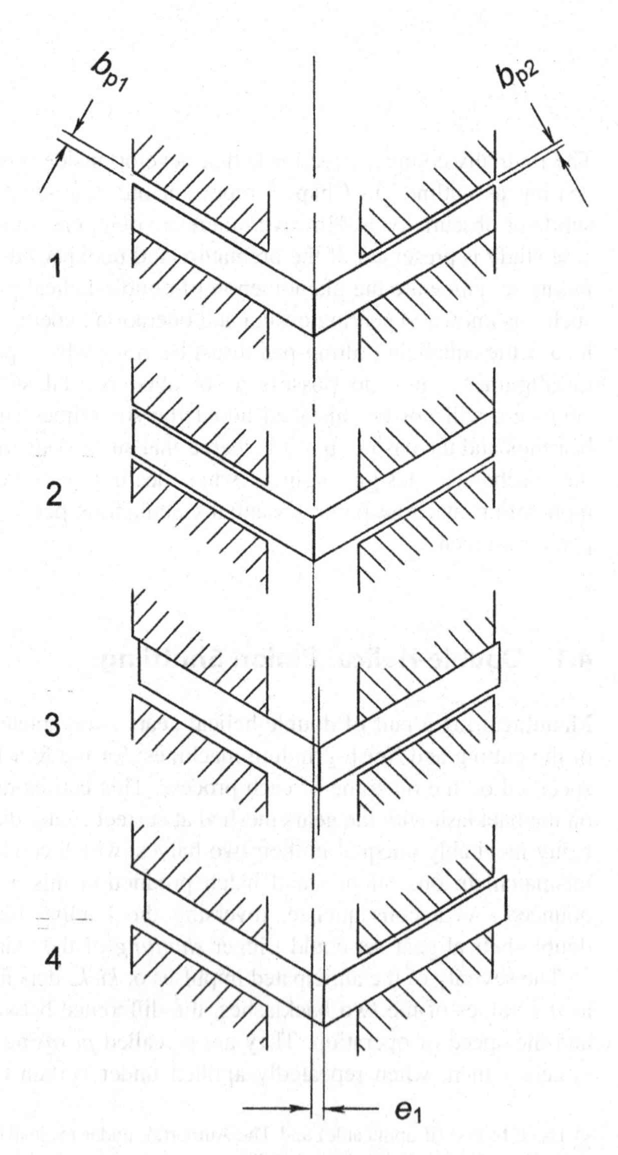

Fig. 4.1 Successive possible relative positions of double-helical teeth of a gear pair upon torque reversal

switched off or the pinion driving torque is reversed, even momentarily. The pinion and gear pitch cylinders are developed into the pitch plane of a conjugate rack. The exaggerated perpendicular distance between the coast flanks is the pitch-plane backlash b_p, its value in the half gear set to the left is assumed larger than in the other half gear set.

Position 1: the pinion drives downwards, centered in the gear tooth flanks; in a reference position. Position 2: driving power is switched off, or torque reverses, the pinion coast side with smaller backlash hits the gear flank first, the pinion receives an impulse to the right, the gear receives the same to the left. Position 3: in slow operation, the pinion reaches the centered position on the coast flanks. Position 4: in a faster process the impulse could make the pinion overshoot to the furthermost possible position to the right by the amount of unilateral end play e_1, the flanks become in diametrically opposite contact. The impulse and its reaction would be smaller than in position 2. In a reversed direction of operation another unilateral end play e_2 would result, and the summation of both is called the (total) end play.

4.1.2 End Play of Unloaded Double-Helical Pinions

In a speed reducing (or increasing) double-helical gear set the gear wheel is usually axially restrained and the pinion is left floating in order to center its two sets of drive flanks in the two sets of the gear-driven flanks, for equal load sharing. The backlash will reside in the coast side of both half gear sets; between the coast flanks, and when the loading direction reverses the drive and coast flanks exchange function. Under no load or at standstill the pinion could be pushed axially in either direction from the centered position to bear on one each of drive and coast flanks. The sum of the two limited displacements is called end play, or end float.

Figure 4.2 depicts a double-helical pinion tooth centered in a gear tooth space. In an end view the normal backlash b_n is seen, which parameter is specified and could directly be feeler gauged, and is of constant magnitude between involute helicoidal flanks regardless of the angle of rotation of the gears. The perpendicular distance between the coast flanks is the pitch-plane backlash b_p,

$$b_p = b_n / \cos \varphi_n$$

The distance between the coast flanks, measured parallel to the other half toothings, could be defined as the double-helical or herringbone backlash b_h; it gives the oblique distance traveled by the pinion, maintaining contact with the flanks in the *other* half, till its drive and coast flanks in the first half switch functions,

$$b_h = b_p / \sin 2\psi$$

The unilateral end play is thus given by

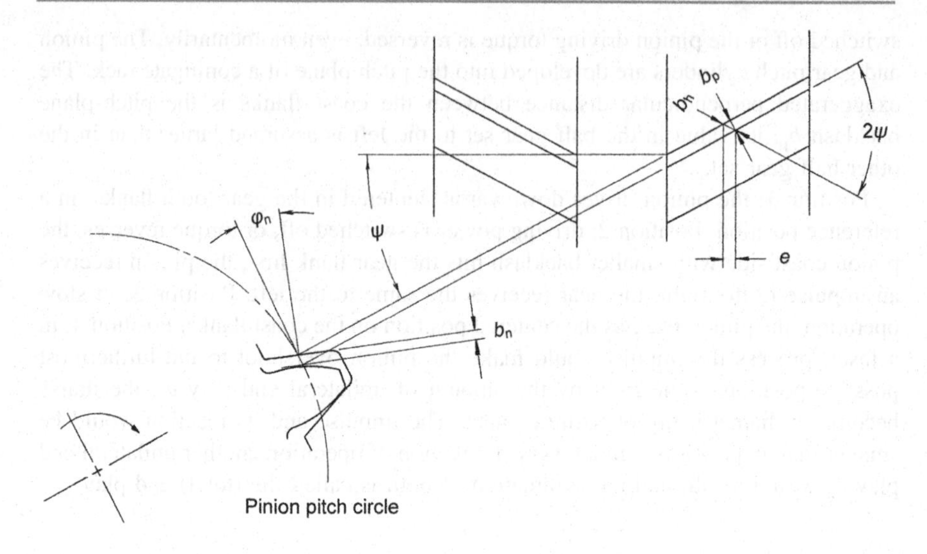

Fig. 4.2 Pinion tooth centered in a gear tooth space in a double-helical pair showing the normal backlash b_n, pitch-plane backlash b_p, double-helical backlash b_h, and the end play e

$$e = b_h \cos\psi = b_n/(2\sin\psi\cos\varphi_n)$$

4.1.3 Case Study

Consequences of double-helical pinion shuttling

System

A single-stage, double-helical gear reducer is used in an oilfield facility to drive a gas compressor from an electric motor. The system is shown schematically in Fig. 4.3. Both shafts are supported in journal bearings followed by an oil thrower on the shaft-terminal side. The output shaft is restrained at its free end by a double-acting, tilting-pad thrust bearing, while the input shaft is axially floating.

Problem Statement

After a few years of operation, the rotating and fixed parts of the oil thrower (slinger) on the input pinion shaft rubbed together and burnt.

Investigation Finding

It was found that the pinion shaft was axially dislocated toward the electric motor (within its double gear coupling) by an amount much in excess of the end play of double-helical gearing. The reason is that the gear wheel itself was found displaced on its shaft by the same amount away from the abutment/shoulder; it lacked any axial retention means. This case was briefly described by Arafa (2006).

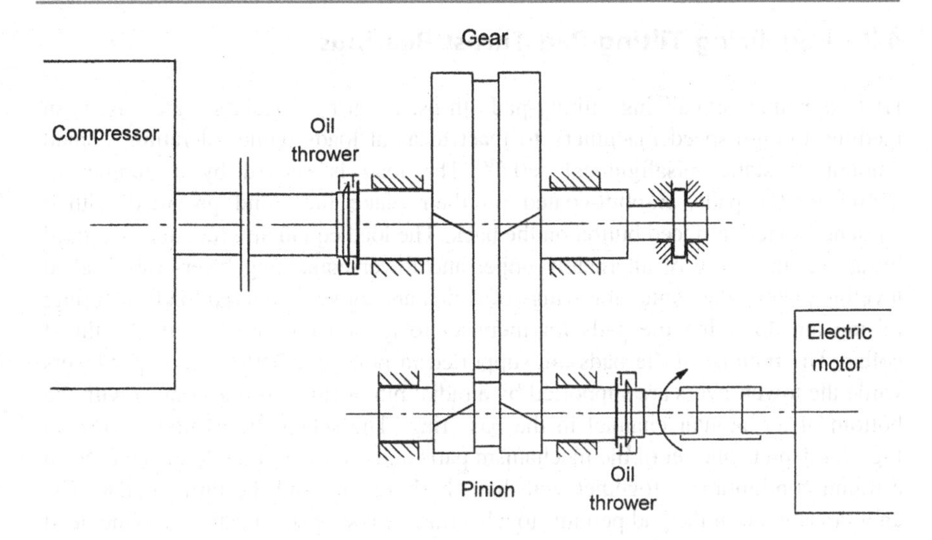

Fig. 4.3 Single-stage double-helical gear reducer schematic

Data, Measurements, Conclusion, and Remedial Action

Normal module $m_n = 6$ mm

Normal pressure angle $\varphi_n = 20°$

Helix angle ψ = arcsin 15/32, one of the recommended values in the German standard DIN 3978 (1976), which was used before the advent of digitally controlled gear manufacturing equipment.

Specified limits on the normal backlash (in the original drawing): 0.216 / 0.368 mm.

Feeler-gauging the normal backlash gave $b_{n1} = 0.36$ mm and $b_{bn2} = 0.26$ mm; difference being 0.1 mm.

Measured end play of the pinion shaft $e_{total} = 0.70$ mm, which is also confirmed by $e_{total} = e_1 + e_2 = (b_{n1} + b_{n2})/(2 \sin \psi \cos \varphi_n)$

The marked difference between the backlashes in the two half gear sets should have imparted the pinion—upon every momentary torque reversal possibly due to torque transients when switching-off the electric motor—a momentum that produced heavy-enough axial impulses on the gear wheel to loosen off its light push-fit (unspecified in the drawing) and wander away from the shaft shoulder. The problem was rectified by assembling the gear wheel on its shaft applying an industrial adhesive specified for retaining cylindrical fits. The damaged oil slinger was also replaced.

Experience Gained from this Case Study

Attention should be paid to specifying the backlash in double-helical gearing; tighter-than-customary tolerances on the backlash should be specified in order for the resulting difference or mismatch to be a minimum. The pinion, being axially floating, requires that the gear be robustly retained.

4.2 Equalizing Tilting-Pad Thrust Bearings

Hydrodynamic equalizing tilting-pad thrust bearings (ETBs) are used in medium-to-high-speed machinery to react to axial loads while tolerating a small amount of static misalignment $\approx 0.5°$. The load is carried by a number of annular-sector pads, Babbitt-coated on their plane face, and provided with a hardened spherical-faced button on the back. The load equalizing function is carried by an annular array of alternately upper and lower equalizing levers (also called leveling plates), the contacting wings of which are convexly curved to allow tilting; raising and lowering the pads for them all to remain conformal with the thrust collar. The buttons of the pads are supported in point contact on the upper levers while the lower levers are supported by a radial pin or ridge in line contact with the bottom of an annular channel in the base ring. The schematic exploded view in Fig. 4.4 depicts one set of the mechanism parts and interfaces that duplicate to form a round configuration, together with half the base ring and the thrust collar. The angular subtend of the pad pertains to a bearing with six pads. Features are included to prevent radial displacement, wandering about, and yawing of the components. The bearing mechanism thus consists of parts that are loosely assembled within the base ring and retained relative to each other and to the base ring through several contacts; simplified in the drawing as salient points. The equalizing levers are shown supported on one another through two points at one end and one point at the other to emulate their ability to tilt.

4.2.1 Pivotal Stiffness of ETBs

Regardless of the two-DOF capability of omnidirectional tilting of the thrust collar, closer inspection of the mechanism reveals that it conceals an additional, independent mobility; the lower and upper sets of levers could assume arbitrary equal tilts in opposite directions, each set in unison, without tilting the thrust collar. Therefore, in order for the bearing parts not to flutter in operation or even assume a hard-over position, the mechanism should be so designed as to be in stable rather than in indifferent equilibrium. The wing contacting surfaces must be so profiled as to raise the pads slightly, should the levers tilt as described, imparting axial *heave* to the thrust collar. As a corollary, when the shaft is slightly misaligned, the different tilting angles of the levers commensurate with maintaining the pads flush with the thrust collar should also impart a heave to the latter. Applying the principle of virtual work, the bearing should thus impose a certain amount of restoring moment on the shaft when misaligned; it should have pivotal stiffness as if to act as a flexure-spring support. And this is an obscure source of a *parasitic* rotating bending moment on the shaft when misaligned. Should the shaft end within the ETB be too tightly sized as not to cope with such a moment, especially with a large angular misalignment, then it could fail in fatigue (see Sect. 6.4). However, bearing manufacturers do not provide data on this flexural stiffness, nor do they mention it.

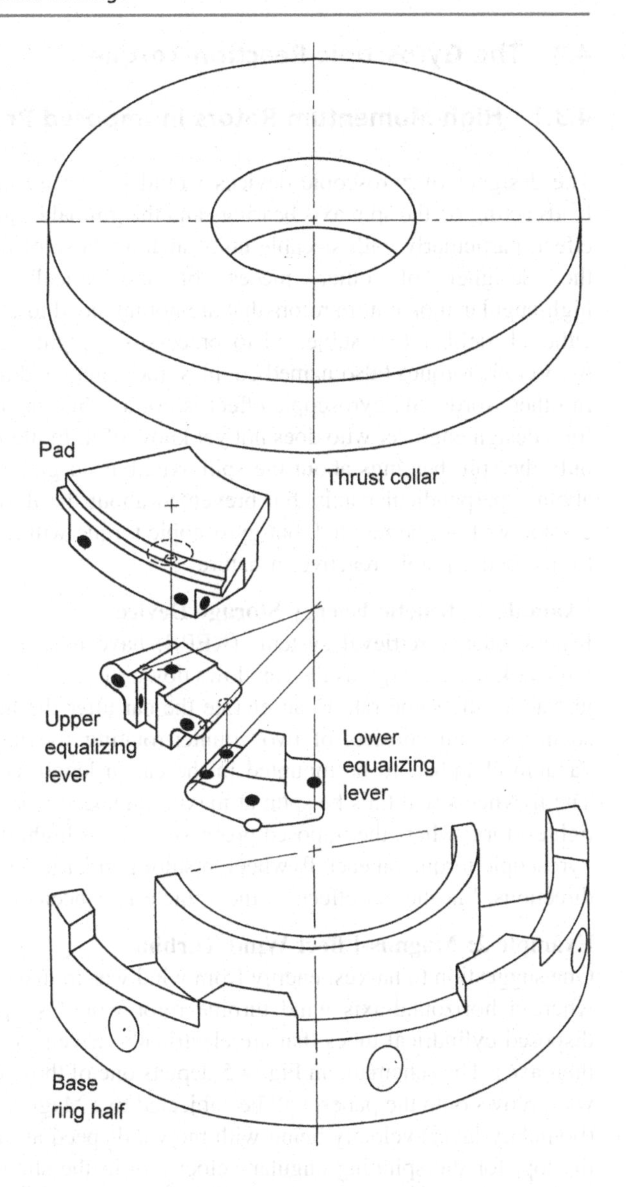

Fig. 4.4 Equalizing tilting-pad thrust bearing in a partial schematic exploded view. The spots establish kinematic equivalence to the contacts

The one and only evidence that ETBs require a moment to tilt under load, viz. to have pivotal stiffness, was provided by Gardner (1985) in his experimental comparison of an industry-standard bearing with a (then) newly proposed design regarding such restoring moment characteristics. The tests were conducted on a static-loading test rig, without the bearings in operation, but the results were graphically presented in a rather sketchy way.

4.3 The Gyroscopic Reaction Torque

4.3.1 High-Momentum Rotors in Imposed Precession

The designer of gyroscopic devices would not be mistaken about the substantial loads acting on the spin axis bearings and the gimbal bearings due to the gyroscopic effect, particularly with sizeable units such as those of marine gyrostabilizers. But the designer of other pieces of mechanical equipment that include high-angular-momentum rotors that are not intended to act or appear as gyroscopes, although deliberately subjected to precession, might well be mistaken about the gyroscopic torques (also named couples; moments) and the bearing loads involved. In other words, the gyroscopic effect is not a subtle or obscure load source except for a design engineer who does not yet know of it. In these devices there are usually only the spin bearings about the spin axis, but no gimbals. Precession is imposed about a perpendicular axis, but prevented about the third perpendicular axis. Precession will not be resisted, but gyroscopic torque will act about that third axis; the torque being purely reactive in nature.

Example 1: Kinetic Energy Storage Device
Kinetic energy retrieval systems (KERS) have been proposed for racing cars to retrieve kinetic energy as the car slows down to negotiate a turn, then to release it at just about the same rate to accelerate the car after the turn. The storage device of such a system consists of two counter-rotating flywheels on parallel axes in a vacuum chamber, to be mounted in the car in lengthwise or transverse direction. The flywheels will thus be spun at maximum speed when the car is negotiating the tightest turn; where the imposed precession rate is highest. This gives rise to a large gyroscopic torque on each flywheel, resulting in large bearing loads in the up/down directions, but the net effect on the vehicle is canceled.

Example 2: Magnus-Effect Wind Turbine
One suggestion to harvest energy from wind was to make use of the Magnus effect, where a horizontal-axis wind turbine rotor would support a number of radially disposed cylindrical tubes that are electrically driven from the inside to spin about their axes. The schematic in Fig. 4.5 depicts one of those cylinders which, when the wind blows onto the paper, will be subjected to a Magnus force F due to the surface (boundary layer) velocity being with the wind speed at the bottom and against it at the top, for the spinning angular velocity ω in the shown direction. For effective energy harvesting the cylinders should be of substantial dimensions, hence substantial mass moment of inertia J, be spun at high speed, and precess at a sensible rate Ω for generating power. This could result in a prohibitive gyroscopic torque $T_g = J\omega\Omega$ to act on the cylinder, the reaction of which tends to bend its supports backwards, unduly loading the bearings, and making a sound mechanical design unfeasible. Negligence or ignorance of the gyroscopic action is considered a design pitfall.

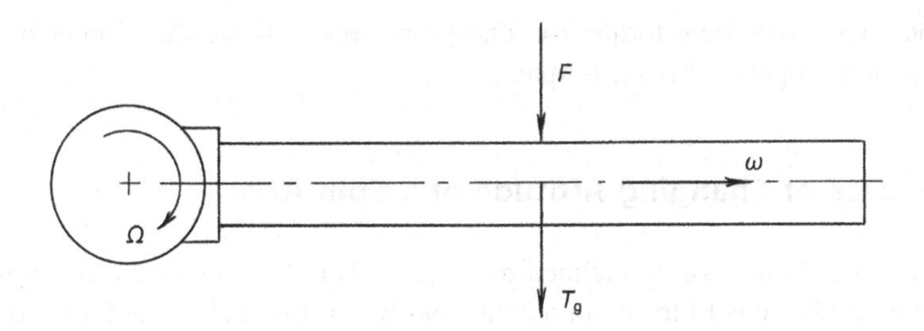

Fig. 4.5 Cylindrical tube, spinning to be subjected to Magnus force as wind blows by

4.3.2 Worked Example

In four-engine aircraft having each two propellers on one wing, counter-rotating, each wing will be subjected to two equal and oppositely acting gyroscopic torques to the right and left when the aircraft pitches (nose up/down), or acting up and down when it changes course (yaws). But the net effect on each wing and on the whole aircraft will be zero. This configuration is shown in Fig. 4.6. Compare the gyroscopic torque with the propeller driving torque when the aircraft pitches at a rate of 1 rad/s

Assumed Propeller Data

Number of propeller blades = 8

Propeller tip diameter d = 5.3 m

Total propeller mass m = 360 kg

Maximum rotational speed n_{max} = 850 rpm

Maximum engine power P_{max} = 8,100 kW

Calculation

1. Maximum propeller tip speed: $v_{max} = \pi d n_{max}/60$ = 236 m/s ≈ 0.76 Mach
2. To calculate the propeller mass moment of inertia the blades will be approximated to uniform beams of constant linear density, so that

   $J = md^2/12 \approx 840\,\text{kg.m}^2$

   $\omega_{max} = 2\pi n_{max}/60 = 89\ \text{rad/s}$

   $T_{max} = J\omega_{max}\Omega = 74{,}760\text{N.m}$
3. Propeller torque $T_P = P_{max}/\omega_{max} \approx 91{,}000$ N.m

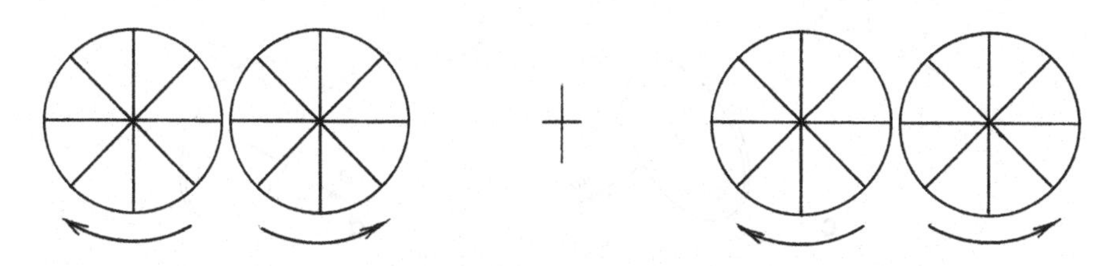

Fig. 4.6 Drawing for the Worked Example 4.3.2

Hence, the gyroscopic torque (on one propeller) is of the same order of magnitude as the propeller driving torque.

4.3.3 Rate of Changing Attitude of a Spin Axis

The two general cases of an inclined rotor axis when it sways or nutates around a fixed intersecting axis to trace out a cone, and when it rotates around a fixed skew axis to trace out a hyperboloid of one sheet, in both cases maintaining a constant inclination angle κ, are depicted in Fig. 4.7a and b, respectively. In both cases the relation between the angular rate of changing attitude Ω (the rate of precession) of the spin axis and the revolutions per second (n) of the rotor around the fixed axis is given by $\Omega = 2\pi n \sin \kappa$.

The first case is commonly encountered in practice, such as with (older) grinding machines with dished grinding wheels for finishing spiral bevel gears like that suggested by Shlesinger and Durkan (1931), in which the grinding wheel axis is made to nutate in a cone, while the second is a rare case. The two special cases of precession of a rotor axis when the conical and the hyberboloidal envelops are flattened out to $\kappa = 0$ are shown in Fig. 4.7c and d; the mostly encountered cases in machinery, where $\Omega = 2\pi n$.

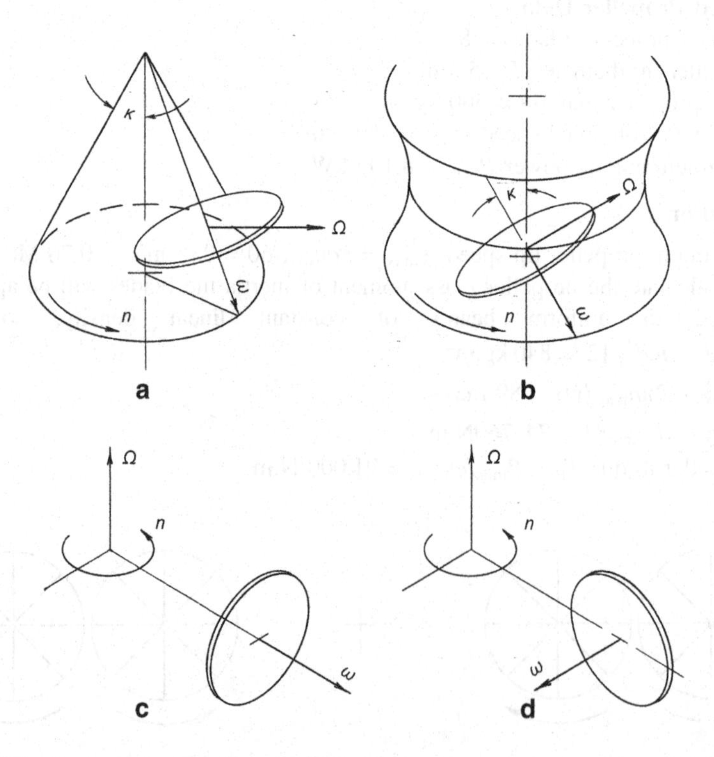

Fig. 4.7 A spinning and precessing rotor of an axis that traces out **a** cone, **b** hyperboloid of one sheet, **c** and **d** plane normal to the precession axis

4.4 Wobble Plate Mechanics

The general case of a rotor spinning on its symmetry axis that nutates to trace out a right circular cone is again posed here in isometric perception in Fig. 4.8, with the rotor centroid placed at the apex of the cone, which is the origin of a fixed Cartesian frame. The rotor/plate will thus be seen to wobble rather than to sway around. The cone axis is made coincident with the *y*-axis, along which the nutation velocity vector ω_n resides. The *y*-*z*-plane contains the momentary spin velocity vector ω_s substantially in the same direction, and separated from the ω_n vector by the semi-cone angle κ, also called the wobble angle. The rotor is thus tilted from the *x*-*z*-plane by κ. The precession velocity vector ω_p will be perpendicular to the spin vector, lying along the rotor diameter momentarily in the *y*-*z*-plane, and its magnitude will be given (according to Sect. 4.3.3 and Fig. 4.7a) by

$$\omega_p = \omega_n \sin \kappa$$

As the spin vector nutates about the *y*-axis, the precession vector will also nutate about the same axis, to trace out an obtuse (shallow) cone of a semi-cone angle of $(90° - \kappa)$. The rate of change of attitude of the precession vector will thus be $\omega_n \cos \kappa$, along the spin axis.

Additionally, the maximum transverse angular acceleration of the plate, momentarily directed along the *x*-axis, will amount to

$$\alpha_P = \omega_n{}^2 \sin \kappa \cos \kappa$$

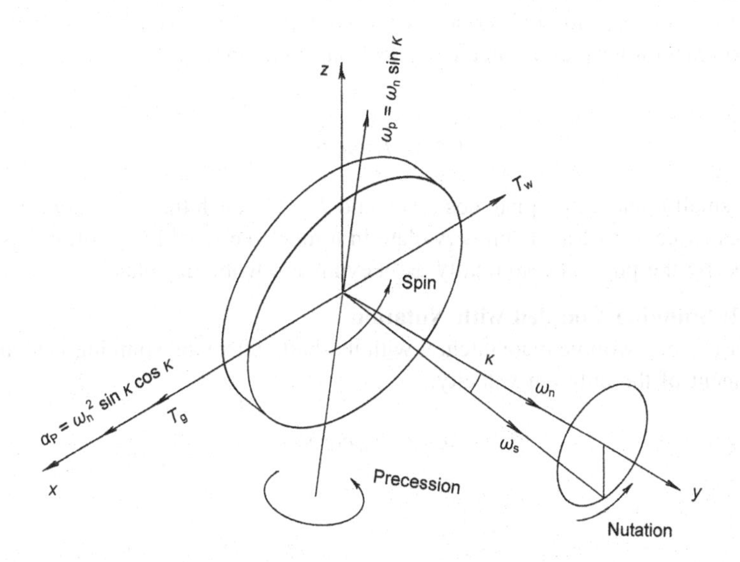

Fig. 4.8 A spinning rotor that nutates or wobbles for its axis to trace out a right circular cone

(The same result is obtained if the two mutually orthogonal vectors ω_p and its rate of change of attitude were multiplied into one another).

4.4.1 Reaction Torque of a Wobble Plate

With J as the mass moment of inertia of the plate about its symmetry axis, the gyroscopic torque due to precession of the spin vector is given by

$$T_g = J\omega_s\omega_p = J\omega_s\omega_n \sin\kappa$$

The gyroscopic torque required to cause precession lies momentarily along the negative x-axis and that reactive torque under imposed precession will be along the positive x-axis; of a restoring nature that tends to decrease the cone angle.

With J_{xx} as the transverse mass moment of inertia of the plate about its central diameter, the inertia reaction torque due to rotor wobbling is given by

$$T_w = J_{xx}\alpha_P = J_{xx}\omega_n^2 \sin\kappa\cos\kappa$$

This torque lies momentarily along the negative x-axis, it is of a decentering nature; tending to increase the wobble angle. It is this torque component that had been neglected in comparison with the gyroscopic torque when dealing with fast-spinning, slowly nutating rotors. Several distinct cases to derive from the foregoing analysis follow.

Case 1: Spinning Decoupled from Nutation

Torque-free motion (inertial motion) of a wobbling plate could be obtained when the two reaction torques T_g and T_w cancel out, implying that

$$J\omega_s = J_{xx}\omega_n \cos\kappa$$

For small κ and a thin plate $\cos\kappa = 1$ and $J = 2J_{xx}$ such that $\omega_n = 2\omega_s$; the plate wobbles twice as fast as it spins. A plate in torque-free wobbling motion has been branded by the physics community as "Feynman's wobbling plate."

Case 2: Spinning Coupled with Nutation

This case is of a wobble plate integral with its shaft, where the spinning velocity is a component of the nutation velocity;

$$\omega_s = \omega_n \cos\kappa$$

A net restoring torque will be present, which amounts to

$$\begin{aligned} T &= T_g - T_w \\ &= (J - J_{xx})\omega_n{}^2 \sin\kappa\cos\kappa \end{aligned} \tag{4.1}$$

This torque will reduce to $J_{xx}\omega_n^2 \sin \kappa \cos \kappa$ for a thin plate. The same result could alternatively be obtained by integrating the effect of centrifugal force on the mass elements of a thin plate when mounted at an incline to its rotating shaft; to wobble with it.

Case 3: Non-rotating Wobble plate

No gyroscopic torque exists, and the wobble plate will only exert a decentering torque T_w. This suggests that when such a non-rotating wobble plate is supported on bearings on a coupled wobble plate then dynamic balance could be achieved when the transverse mass moment of inertia of the non-rotating wobble plate is made equal $(J - J_{xx})$ of the coupled wobble plate.

This result is useful for designing wobble plate machinery to be approximately dynamically balanced for increasing the operating speed limit and, hence, for improving the performance density. An example is shown in Fig. 4.9 of the *mechanical end* of a hydraulic axial-piston wobble plate motor with eleven pistons and a wobble plate tilt angle of about 18° (see also Fig. 11.1). Two identical thrust ball bearings of the extra heavy series are used; one being mounted on the shaft wobbling abutment for the non-rotating, face-centered ring to generate the sinusoidal stroke motion of the pistons without much sliding on their conical ends and to take their heavy axial reactions, while the other bearing routes the same axial reaction to the casing. Spring-loaded pistons are used to cope with their oscillating inertia forces, and to keep the bearing parts in place when not in operation. Part of the shaft and the bearing ring fitted thereon represent an integral or coupled wobble plate, the other bearing ring is a non-rotating wobble plate that partially counteracts the gyroscopic reaction torque of the first part, while the set of balls with their cage rotate at half the wobble rate; an intermediate case thus expected to be self-balanced (except for their small eccentricity). This is because the ball cage actually emulates the Feynman's wobbling plate just discussed; a rare encounter in mechanisms.

4.5 The Isoinertial Solid Cylindrical Rotor

The net restoring torque on a wobble plate integral with its (mass-less) shaft is given by Eq. 4.1, wherein ω_n is the nutation velocity; the shaft angular velocity proper, denoted ω hereinafter. Therefore, complete dynamic balance of a solid, right circular cylindrical rotor could be achieved when $J = J_{xx}$; the rotor thus having the same principal moment of inertia about any axis passing through the centroid, when rotated about any of those axes. Any set of orthogonal axes through the

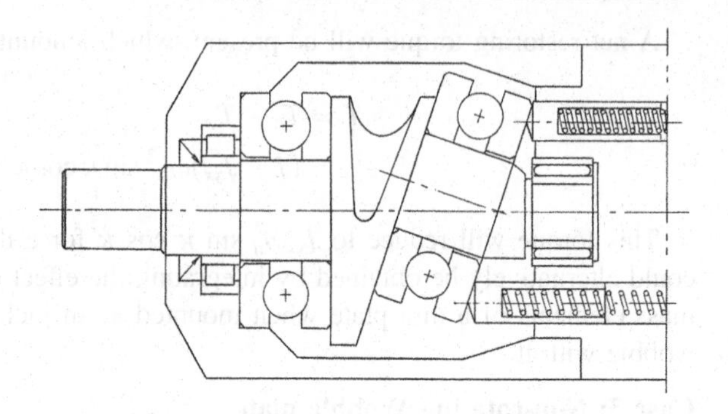

Fig. 4.9 Hydraulic axial-piston motor of the wobble plate type

centroid, regardless of their slant, will serve as principal axes too. Therefore, with a rotor material density ρ and for $T = 0$,

$$\pi\rho LR^4/2 = \pi\rho LR^2\left(3R^2 + L^2\right)/12$$

This gives the condition $L/R = \sqrt{3}$ for any value of κ; for the rotor to rotate about an axis of any inclination to its principal axis, passing through the centroid. Such a rotor with identical principal moments of inertia is also said to have *isotropic inertia*; to be *isoinertial* (just like a sphere or a cube). These rare characteristics of the $\sqrt{3}$-ratio cylindrical rotor may be helpful in designing inherently/directly balanced components. But the actual shaft of the rotor—to either side or both—will have a slanted face, which feature should be taken into consideration in unbalance calculations.

An alternative derivation of the required L/R ratio of an isoinertial solid cylinder could be based on the condition that the summation of the moments, about a transverse axis through the centroid, of the centrifugal force components of all the mass elements, in planes parallel to the one that momentarily contains the rotation axis and the geometric axis, will be set to zero. Figure 4.10 in two views shows a mass element dm in this setting, and the necessary geometry. (Radius r from the cylinder axis to the mass element and its instantaneous position θ do not appear to their exact measure in the projection). The mass element rotates about the ω-axis at an instantaneous radius w of which the vertical projection is v, and the arm length to the transverse axis through the centroid is a. Therefore,

$$\begin{aligned}
\mathrm{d}m &= \rho r\,\mathrm{d}\theta\,\mathrm{d}r\,\mathrm{d}z \\
\mathrm{d}F_\mathrm{v} &= \omega^2 v\,\mathrm{d}m \\
v &= (r\cos\theta - z\tan\kappa)\cos\kappa \\
\mathrm{d}M &= a\,\mathrm{d}F_\mathrm{v} \\
a &= z/\cos\kappa + (r\cos\theta - z\tan\kappa)\sin\kappa
\end{aligned}$$

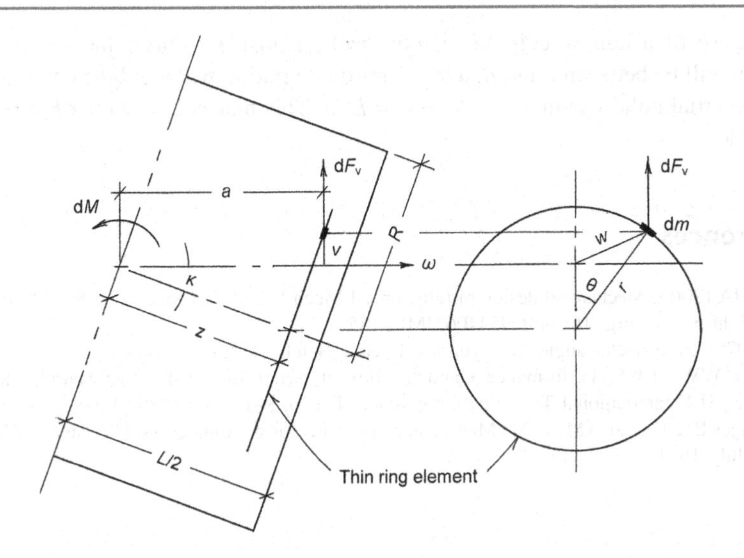

Fig. 4.10 Derivation of the condition for an isoinertial solid cylinder

Therefore, substituting and simplifying,

$$\mathrm{d}M = \rho\omega^2\left[(1 - 2\sin^2\kappa)r^2 z\cos\theta + (\sin\kappa\cos\kappa)\left(r^3\cos^2\theta - rz^2\right)\right]\mathrm{d}\theta\,\mathrm{d}r\,\mathrm{d}z$$

Performing triple integration, first over a thin ring element (subscript $_\mathrm{R}$) at a distance z along the cylinder axis (although the mass element rotates about the ω-axis) with θ between 0 and 2π, keeping r and z constant, to obtain the moment due to the thin ring element

$$\mathrm{d}M_\mathrm{R} = \rho\omega^2(\sin\kappa\cos\kappa)\left(\pi r^3 - 2\pi rz^2\right)\mathrm{d}r\,\mathrm{d}z$$

then over a thin disc (subscript $_\mathrm{D}$) that contains the ring, with r between 0 and R, keeping z constant

$$\mathrm{d}M_\mathrm{D} = \rho\omega^2(\sin\kappa\cos\kappa)\left(\pi R^4/4 - \pi R^2 z^2\right)\mathrm{d}z$$

then over the whole rotor with z between $-L/2$ and $+L/2$

$$M = (\pi/4)\rho\omega^2(\sin\kappa\cos\kappa)R^2 L\left(R^2 - L^2/3\right)$$

This moment is zero when the ratio $L/R = \sqrt{3}$.

In case of a hollow cylindrical rotor with an inside radius r the second integration will be between r and R, and the third integration will yield the condition for an isoinertial hollow cylinder as $R^2 + r^2 = L^2/3$. This makes $L = D$ for $r/R = 0.6$, for example.

References

Arafa HA (2006) Mechanical design pitfalls. Proc I Mech E Part C, J Mech Eng Sci 220(6):887–899. https://doi.org/10.1243/09544062JMES185

DIN 3978 (1976) Helix angles for cylindrical gears. Beuth, Berlin

Gardner WW (1985) Performance characteristics of two tilting pad thrust bearing designs. Proc JSLE International Tribology Conference, Tokyo, 8–10 July 1985:61–66

Shlesinger BE, Durkan TM (1931) Method and apparatus for grinding gears. US Patent 1,815,336, 21 July 1931

Interface Sliding, Loading, and Wear 5

The durability of mechanical equipment depends much on the design measures taken to reduce wear, which is function of load and sliding speed. Minimizing unnecessary reactive loads has already been dealt with in Chaps. 2 and 3. Defined here will be the so-called *interface sliding volume* and *interface sliding area* per cycle of operation, which are included within those design parameters used to make a comparative assessment of various design concepts regarding durability. Several design examples are included to highlight this measure. Application of the principles of self-reinforcement (viz. loading upon demand) as well as of homogeneous-wear interface design are also addressed in the context of durability enhancement. Homogeneous loading and wear is also an issue of importance to the design of ball screw nut systems, which has hitherto not been satisfactorily treated. This will be analyzed in detail to assess the "optimum" number of ball track turns inside the nut.

5.1 Interface Sliding Per Cycle

Two definitions of the so-called *interface sliding* are applicable to the various contact geometries between sliding mating surfaces:

1. The interface sliding volume (mm^3); the conformal contact area times its relative sliding displacement.
2. The interface sliding area (mm^2); the length of a contact line times its sliding displacement, which is not necessarily perpendicular to its direction.

If the sliding velocity varies throughout the contact area or along the contact line, then integration should give the sought average value.

H. A. Arafa, *Design for Durability and Performance Density*,
https://doi.org/10.1007/978-3-030-56816-0_5

Energy loss and surface material distress due to sliding friction, such as wear and scoring are function of the sliding, loading, and the frictional force at the mobile interface between mating components. Therefore, minimum interface sliding (volume or area) per operating cycle is considered an attribute or a merit criterion in mechanical design; it could become one of the *subordinate* design principles. The heavier the load the more relevant this principle becomes. One operating cycle is taken as a reference for a comparative assessment of different designs. This could be a shaft revolution, a stroke or a full reciprocation, a cycle of actuation and de-actuation, and the like.

Comparison is to be held between geometries of one and the same (basic) machine element that could be styled with different sliding volumes or areas such as pistons for internal combustion engines, or between different machine elements that have the same functional requirement. Examples may be mentioned such as

1. Reciprocating piston engines versus rotary piston engines.
2. Poppet valves (stem inside guide) versus rotary overhead valves.
3. Quad-link couplings versus dual multi-crank couplings.
4. Linkages with revolute joints versus ones with prismatic joints (slides).
5. Gearing (in general) versus worm gear drives.

5.1.1 Pistons of Minimum Skirt Area

Figure 5.1 is a schematic of a piston intended primarily for high-performance internal combustion engines. The piston features a crown that holds two compression rings and one oil control ring, as customary. The piston skirt is such contoured as to leave an adequate bearing area on the major thrust side, and a smaller bearing area on the minor thrust side. This design results in minimum interface sliding volume per cycle, compared to the much more commonly used partial-skirted and full-skirted pistons. In automotive engineering this style piston is referred to as an asymmetrical piston. The contoured skirt area reduces the frictional power loss, allowing higher engine speeds, hence higher power density.

The unsymmetrical mass distribution shifts the center of mass from the piston axis, so that some manufacturer offsets the gudgeon pin bores by the same amount (toward the major thrust side) to eliminate cyclic inertia tilting moments on the piston that would make it rock within the cylinder wall. However, what the piston manufacturer does not mention is that the gas force vector (due to the gas pressure on top of the piston) always acts along the cylinder axis, no matter how the topography of the piston top surface appears, which again produces rocking with an offset gudgeon pin. It would have been better to retain the centrically disposed gudgeon pin and achieve mass balance by putting some more mass on the narrow-skirt side.

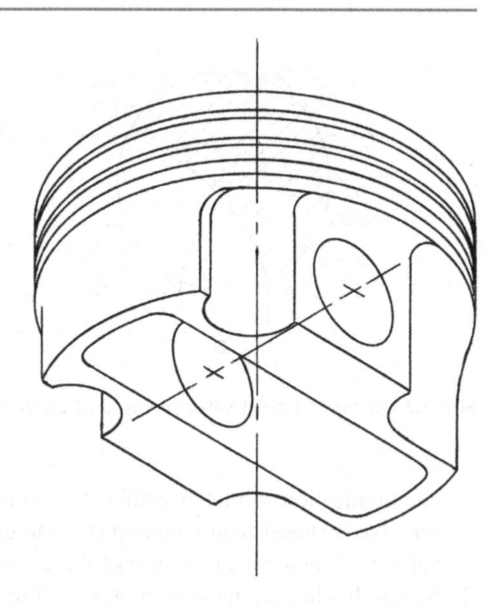

Fig. 5.1 Asymmetrical piston of an internal combustion engine

5.1.2 Engine Valves

The best proven valve for internal combustion engines has always been the poppet valve. The valve head is seated with its 45° chamfer on a hardened valve seat; the larger the gas pressure the tighter the sealing effect; a straightforward application of the principle of self-reinforcement. Performance degradation is readily rectified by lapping the valve on its seat; good maintainability. The valve stem is guided in a valve guide, which is press fitted into the cylinder head and could be replaced when necessary. The interface is being sufficiently lubricated and is not subjected to side thrust. The small interface area times twice the small valve lift result in a very small interface sliding volume per cycle, in this case one four-stroke cycle; two crankshaft revolutions. Typical dimensions for an 80-mm bore engine with two valves per cylinder are a valve stem diameter of 7 mm, valve guide length of 50 mm, and valve lift of 8 mm, which give an interface sliding volume of $\pi \times 7 \times 50 \times 2 \times 8/1000 \approx 17.5\ \text{cm}^3$ for one valve, or 35 cm^3 for the pair of valves, per cylinder.

For over a century—and still presently—inventors have been obsessed by the apparent simplicity and motion uniformity of rotary overhead valves. They covered all possible combinations of spool gas-passage geometry, number of spools (one or two), and number of gas-exchange openings in the combustion chamber (one or two). The suggestions are only known as patents, seldom assigned to any manufacturing company, and having been put in operation, satisfactorily, is doubtful. The suggested gas passage geometries fall into two categories.

1. Spools having quadrant routes, in form of circular segment flats (or slightly concave ones), such as in Fig. 5.2a, either to establish one same gas

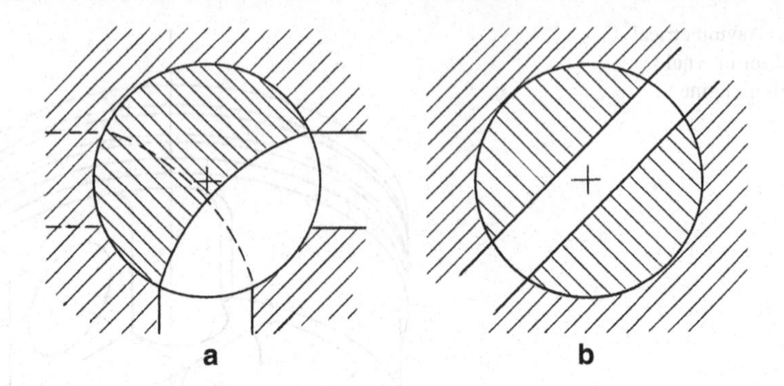

Fig. 5.2 Rotary overhead valve spools, **a** of quadrant routes, **b** of diametrical routes

communication with the combustion chamber every spool turn, or two consecutive (exhaust then intake) gas communications every spool turn. The spool valve will then be driven at half the crankshaft speed.

2. Spools having diametrical routes, in form of slots, to establish the same gas communication twice every spool turn such as in Fig. 5.2b, to be driven at *one quarter* the crankshaft speed.

Further comparative assessment would be out of the scope of the present text. But the basic problem with rotary valves is that the high pressure in the cylinders pushes on the valve against the top half of the sliding surfaces while the valve is rotating, causing excessive friction, wear, and ultimately seizure. Maintenance would imply replacing the long valve spool by an oversize one and reconditioning the long bore; a very difficult task. Maintainability is therefore too poor, and rotary valves do not belong to those well-proven, surviving designs. From another perspective the sliding volume per cycle is too large. There is no industry standard for rotary valves, but a value for an 80-mm bore engine with a cylinder spacing of 100 mm and a valve diameter of 42 mm would amount to $(\pi \times 42)^2 \times 100/1000 \approx 1{,}516\ \text{cm}^3$ per cylinder. This is almost two orders of magnitude larger than for poppet valves.

5.1.3 Couplings for Parallel Offset

The quad-link coupling in Fig. 5.3a consists of two drive discs and one middle disc, linked to either drive disc by two short, parallel couplers. The two sets of couplers are arranged at right angles to one another and are symmetrically disposed about the principal axes. This resembles two planar isograms in series to handle a small parallel offset between two shafts. The coupling could be conceived as a seven-link mechanism. The couplers have bearing bushes of best quality and make only a limited oscillation on the pins, proportional to the shaft offset. The design thus

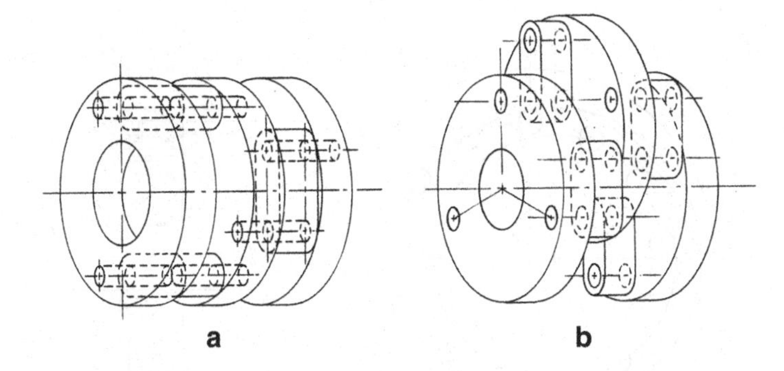

Fig. 5.3 Couplings for parallel offset, **a** quad-link coupling, **b** dual multi-crank coupling

complies with the principle of minimum interface sliding volume per cycle. The coupling is shown in its reference, coaxial position. With a parallel offset the middle disc shuttles between the centers of the two drive discs at twice the rotational frequency; the coupling thus suffers some dynamic unbalance.

The dual multi-crank coupling in Fig. 5.3b consists of three discs connected in series by two sets of three or more parallel cranks. It is intended to cope with parallel shaft offset to a much higher extent than the quad-link coupling. The system is shown in its coaxial position, where the two eccentricities to the middle disc coincide. But this position should be avoided in operation; it imposes a constraint on the discs in the direction of the eccentricity, and manufacturers never depict their product in this position.

With parallel offset the middle disc center adjusts at the apex of an isosceles triangle with the centers of the two shafts. The cranks have bearing bushes of best quality and should be well lubricated. However, this assembly is a *drive* rather than a coupling; the middle disc is driven about its center at the same shaft speed and the bushes rotate on the pins continuously at the same speed. This design does not comply with the principle of minimum interface sliding volume per cycle.

5.1.4 Hydraulic Distributors

Hydraulic axial-piston motors of the wobble plate type (in which the cylinder block is stationary, see Figs. 4.9 and 11.1) can have their distributor (commutator) in form of an orbiting eccentric ring as shown in Fig. 5.4. The tail shaft of the motor carries an eccentric disc, a needle-roller bearing cage, and a ring with wider rim. The face-land radial width of the rim is just a little larger than the cylinder port diameter. As the eccentric ring orbits, needless of rotation, it connects each cylinder to a high-pressure plenum or a low pressure plenum, which are on the inside and the outside of the ring, or vice versa, according to the sense of rotation.

The ring is a telescopic two-L-section component, thus featuring an outside and an inside radial gap, to be loaded in axial direction by the supply pressure acting on

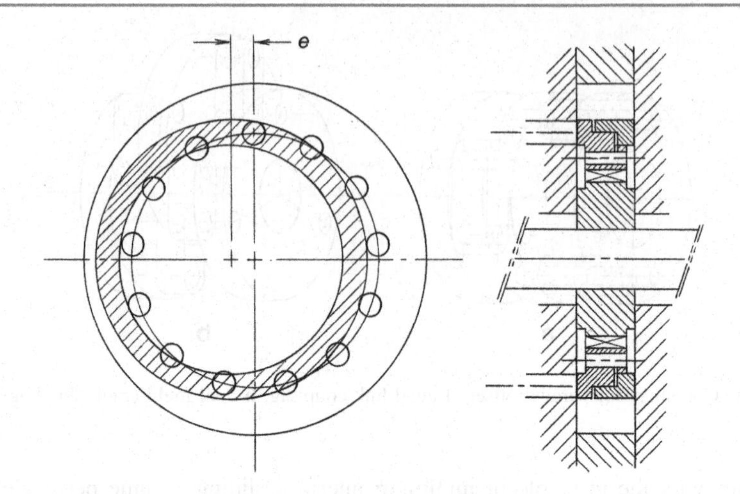

Fig. 5.4 Hydraulic distributor of the orbiting eccentric-ring type

either one according to the pressure prevailing on the outside or the inside. The back face of the ring slides on the casing cover. The eccentric ring is perforated for axial equilibrium and in order to access the inner plenum from the rear. The eccentricity vector is perpendicular to the plane containing the axes of the motor and the wobble plate so that any piston at top or bottom dead center is cut off. In this system covering and uncovering the cylinder ports occurs gradually, providing a flow area proportional to the piston velocity, resulting in smooth running at low speeds and less noise at high speeds. Figure 5.4 shows the conditions in a 13-piston motor. Two design principles are being satisfied by this configuration:

1. Self-reinforcement. The pressing force of the ring face lands is made proportional to the operating pressure, a feature that is autonomously combined with self-adjustment.
2. Minimum interface sliding volume per revolution. The ring only orbits without rotation, so that all the surface elements of the face lands have one and the same sliding speed of ωe, which is much smaller than ωr in case of a rotating valve element of an average radius r.

The same principle is also applied to some types of hydraulic radial-piston motors with five or seven cylinders, where the conduits to the cylinder heads are made to terminate on a transverse plane surface for the eccentric ring to operate on. In this case the eccentricity vector is perpendicular to the crankshaft throw, and the cylinder ports are such contoured as to provide larger flow areas within the same confined radial width of the face land.

Some makes of radial-piston motors have their distributor in form of a cylindrical spool (viz. rotary union), which is in a close running fit with the housing and connected to be driven by the motor shaft, Fig. 5.5. Both the interface area between

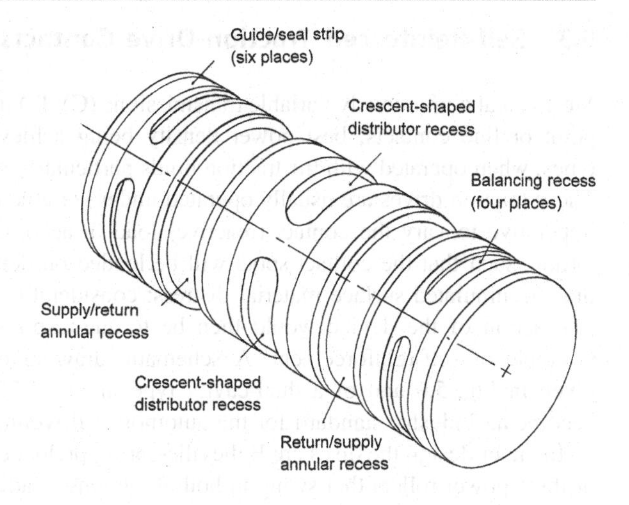

Fig. 5.5 Hydraulic distributor of the cylindrical spool type. Not shown are the internal conduits between the supply/return and the distributor and balancing recesses

spool and housing and the sliding velocity are not small, so that the interface sliding volume per revolution is appreciably higher than with the orbiting eccentric-ring type. In addition, the spool requires four balancing recesses in order not to jam inside the housing by the rotating radial load acting in its middle due to the one high-pressure distributor recess.

5.2 Rolling Versus Sliding Interfaces

Sliding interfaces that operate under sustained full hydrodynamic fluid films are unbeatable regarding durability. One famous example is internal combustion engine journal bearings, where all endeavors to replace them by roller bearings have failed. However, frictional power losses could be, in general, minimized when loaded sliding interfaces that operate under boundary lubrication conditions are replaced by rolling-element ones.

Examples:

1. Ball screws or planetary roller screws versus conventional threaded power screws.
2. Tripod joints versus gear couplings (both being plunging joints).
3. Rzeppa joints versus the once famous tracta joints (famous for their overheating and scoring), both being non-plunging, constant-velocity joints capable of accommodating large operating angles, primarily for front-wheel-drive vehicles.
4. Rolling-element worm gearing versus conventional worm drives (in low-speed operation), such as the long-forgotten globoid worm that drives a wheel with a number of conical or spherical-segment rollers supported on needle bearings over radial stub axles. These have been described by Pekrun (1912) and Schiebel (1913).

5.3 Self-Reinforced Traction-Drive Contacts

Mechanical continuously variable transmissions (CVTs) depend on traction through point or line contacts; best power density being achieved with the half-toroidal types, when operated with the traction fluids particularly developed for the purpose. These traction drives are usually operated under variable torque loading. Thus, it is imperative to vary the contact (reactive) load in accordance with the transmitted torque, such that the contact spots will be loaded on-demand. This minimizes the time-accumulated surface material distress; considerably improving the durability. The action of the device would then be torque-responsive; applying the design principle of self-reinforcement. A schematic drawing of a single-cavity type is given in Fig. 5.6 and of a dual-cavity type in Fig. 5.7, the latter having almost become an industry standard for the automotive driveline.

In either design the discs are Belleville-spring preloaded against the set(s) of two or three power rollers that swing in both directions to adjust the transmission ratio, and the driving torque is applied through a roller–ramp device such as to produce

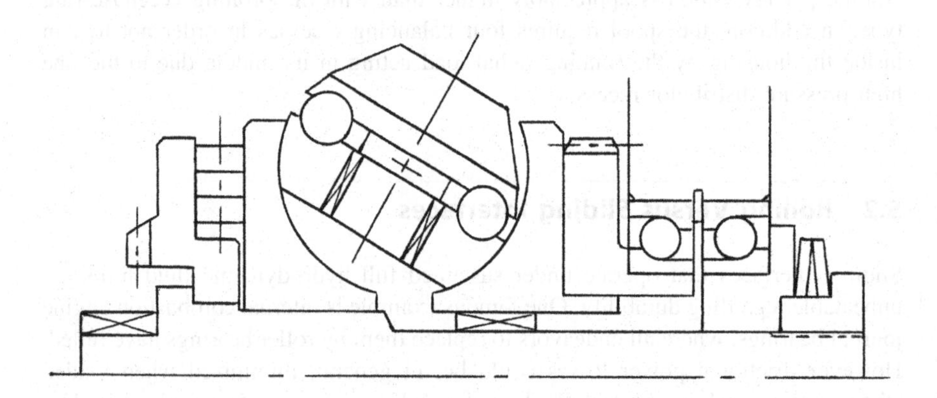

Fig. 5.6 Half-toroidal, single-cavity CVT

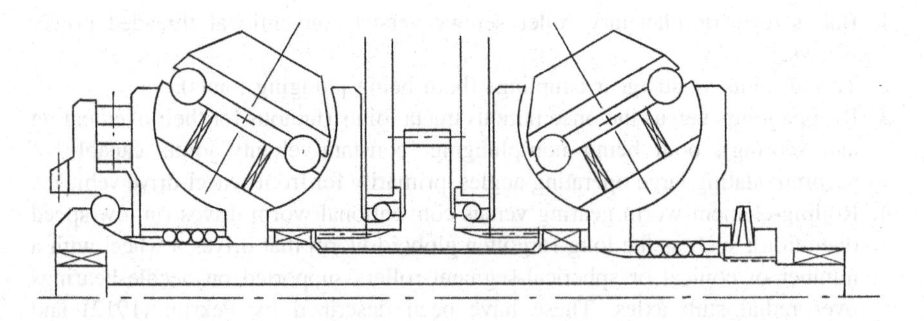

Fig. 5.7 Half-toroidal, dual-cavity CVT

the additional, main proportional loading. This device consists of the back side of one of the outer discs and an input flange, both having four ramps on their faces and being separated by cylindrical rollers guided in a cage to remain radial. The input flange is driven by an input shaft through a dog clutch to remain axially unconstrained.

In single-cavity half-toroidal CVTs the two (unequal) axial reactions of the discs are routed to the casing, by necessity through two angular-contact ball bearings. The bearing of the input drive disc receives its axial load through the (torque-free) shaft in tension, while the other bearing is in direct abutment with its disc. It is evident that these two angular-contact ball bearings operate under high loads at high speeds and should therefore be of ample size, for them not to be the first item to fail in fatigue. They are source of an additional, frictional power loss. The casing material extending between the two bearing outer-race abutments should also be of sufficiently high stiffness such as not to affect the axial separation of the two discs. This is being achieved in some designs, such as shown, by disposing the two bearings back-to-back on one side.

In dual-cavity half-toroidal CVTs the driving torque is applied to the two outer discs, which are ball splined on the hollow shaft such as to undergo micrometric axial displacements. The input flange pulls the shaft through an angular-contact ball bearing that only allows the relative angular freedom required for actuating the device—it does not actually rotate under load. Angular-contact ball bearings running under high load are dispensed with, since the two equal axial reactions on the middle discs directly cancel out, and the other two equal reactions on the outer discs are balanced through the shaft in tension. This design therefore complies with the principle of *directly balanced reactions*, rather than *balancing over bearings*. Output power is taken from the two inner toroidal discs by the gear in-between. The casing has only to support the assembly in radial direction and to retain it in axial direction, under no load. Durability of the CVT will thus neither be affected by such distressed ball bearings nor by the casing compliance.

5.4 Homogeneous-Wear Interface Design

Mobile interfaces of appropriate material pairings are subjected to normal wear due to the relative sliding and/or rolling under load, leaving smooth surfaces without scores. These interfaces should be such configured as to exhibit as homogeneous-wear pattern as possible, in order to preserve their functionality over longer durations of operation, hence to increase the durability.

In case when operation is under unilateral loading, homogeneous-wear interfaces render the components self-adjusting. When under bilateral loads, homogeneous-wear interfaces lend the components to being re-adjustable; for clearance or preload. This implies (infrequent) stopping and performing some fine adjustment; an application of one of the principles of maintainability. Examples include taper-cut spur or helical gears, double nuts, and spherical joints with a spherical-segment counterpart.

Some of the ways by which homogeneous wear could be achieved at an interface would qualify or be regarded as design principles, to be added to the already existing repertoire, and are given with examples in the following.

1. Continually varying exposure of interface elements:
   vane pumps and radial-piston pumps with floating stroke ring,
   inverted-bucket tappets with offset cams.
2. Asynchronous pairing of multiple identical interface elements:
   hunting-tooth gear pairs.
3. Gradual null-crossing of points along a cyclically sliding contact line:
   helical gears.
4. Even load distribution on discrete load points:
   ball screws with the nuts of the "optimum" number of track turns.
5. Even load distribution along contact lines:
   self-aligning line contacts such as in curved-tooth gears (see Sect. 14.4).
6. Even load distribution over contact surfaces:
   self-conforming interfaces such as spherical joints.

5.4.1 Continually Varying Exposure of Interface Points

An example of application of this principle is hydraulic variable-displacement vane pumps, having their vane outer edges slide inside a cylindrical stroke ring, which could be controlled to a variable eccentricity to adjust the stroke of the vanes, hence the displacement volume as shown in Fig. 5.8. The vanes are pushed onto the stroke ring for sufficient sealing by centrifugal force and by applying the discharge pressure on their inner edges, at least partially. The stroke ring reacts on a radial thrust pad under the discharge pressure that acts on its inside surface during the discharge half of a revolution. The stroke ring is left rotationally floating; it can rotate at a small yet uncontrolled rate, to evenly distribute the wear over its inside surface. A similar concept had also been adopted in radial-piston pumps with inward compression stroke, where the connecting rods end in slippers that slide inside a floating stroke ring.

Another example is engine valve operating mechanisms of the inverted-bucket-tappet design with an overhead camshaft, where each cam is a little offset from the axis of the bucket tappet to promote a slow rotation of the latter such as to even out wear and maintain good seating, Fig. 5.9.

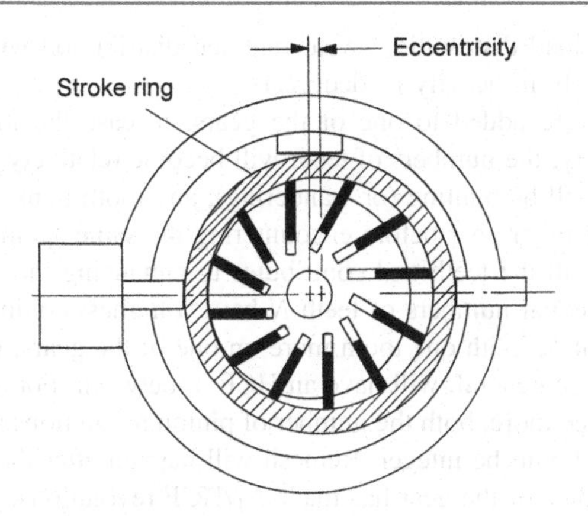

Fig. 5.8 Hydraulic variable-displacement vane pump

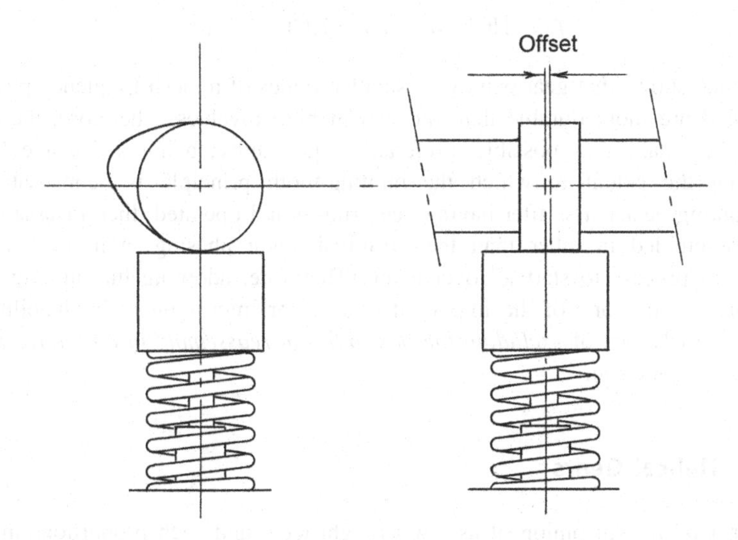

Fig. 5.9 Engine valve with an inverted-bucket tappet

5.4.2 Hunting-Tooth Gear Pairs

In a pair of gears of equal numbers of teeth, each driving flank will be dedicated to a certain driven flank in the mating gear, which it will contact repeatedly every revolution. Both the running-in process and the subsequent operation of the gears will be characterized by each tooth flank being matched to its associated flank regarding inaccuracies, in isolation from all other flanks. The result will be a geometry of each two flanks that is slightly different from the others. Noise generation, tooth load distribution, wear rate and distribution will all be points of concern, particularly in heavily loaded gears.

If one tooth were added to one of the gears, in case the ratio is not strictly required to be unity, the numbers of teeth will become relatively prime (co-prime), and all the teeth will be hunting (or wandering); any tooth flank will contact every tooth flank on the other gear before encountering the same again. This spreads the wear evenly over all the teeth and contributes to enhancing the durability.

Two gears of equal numbers of teeth N have a highest common factor (abbreviated HCF) equal N. With one tooth more on one of the gears, the HCF becomes unity. Gear pairs, in general, will have an HCF in-between. For a pair of reference flanks to mesh once more, both the number of pinion revolutions and the number of gear revolutions should be integer. Remesh will happen after the pinion has made N_G/HCF revolutions, or the gear has made N_P/HCF revolutions, which results in a tooth remesh frequency, or a hunting-tooth frequency

$$f = \text{HCF}\ (n_P/N_G) = \text{HCF}(n_G/N_P)$$

Practice shows that gear pairs with smaller values of remesh frequency perform better and are more durable than those with higher values. Therefore, the HCF should be smallest as possible, preferably equal unity; being a decisive factor regarding the extent to which the hunting-tooth principle is being satisfied. Non-hunting gear pairs, after having been run-in and operated, then disassembled and reassembled in other than their original angle phasing, will need a new running-in process to start all over again. Therefore, adopting the hunting-tooth principle satisfies one of the important criteria for improving maintainability by design, namely that of *excluding the possibility of reassembly in an inadvertently incorrect orientation.*

5.4.3 Helical Gears

A right-hand helical pinion of as few as eight teeth and such proportions that its tooth flanks clearly exhibit their involute helicoidal geometry is shown in Fig. 5.10a. The basic geometry of an involute helicoid is shown in Fig. 5.10b (with a smaller lead angle, to resemble an involute worm flank, rather). This is a ruled developable surface, where its straight generatrix is always tangent to the base cylinder and any transverse section is an involute. Instantaneous contact of the pinion with a meshing helical gear will be along a common generatrix that lies in the plane of action, which plane is tangent to both base cylinders and inclined at the transverse pressure angle to the common tangent plane to both pitch cylinders. One instantaneous contact line is shown on one tooth flank, covering the working depth up to the tooth tip; this being possible due to the rather large helix angle. The contact line on the following tooth flank is shown to lie at a deeper level, to terminate below the tooth tip. As the gears rotate the successive positions of the

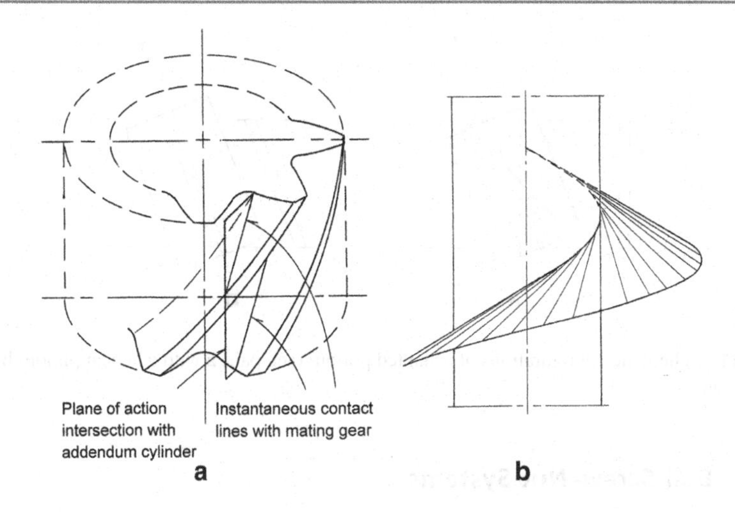

Fig. 5.10 Involute helicoidal surfaces, **a** tooth flanks of a helical pinion of eight teeth and a face advance of 1.5, **b** basic geometry

contact lines in the plane of action make an array of parallel straight lines delimited by the side planes and the addendum cylinder. But the successive positions of the contact line on one tooth flank resemble successive generatrices of the involute helicoid which are not parallel, as seen in Fig. 5.10b.

Contact between helical gears is therefore continually varying to expose interface points at all relative sliding conditions at the same instant, including the contact point on the pitch line, where sliding stops and reverses. Helical gears are therefore known for their superior running-in characteristics that result in better accuracy and surface finish than originally manufactured, and for their favorable subsequent operation regarding quietness. They are more durable than straight spur gears.

Some rack and pinion applications require that the pinion axis be slanted rather than perpendicular to the orientation of the rack. The slant could be achieved by having either a straight-cut rack meshing with a helical pinion, or a *helical* rack meshing with a spur pinion, as shown in Fig. 5.11a and b, respectively. However, a rack is neither spur nor helical regarding the meshing characteristics; in either case it is just a slice of one and the same continuum. The merits of truly helical mesh are only obtained with a helical pinion. But when a spur pinion is meshed with a *helical* rack usual spur mesh action will result, with the known drawbacks. Therefore, this will be regarded as a deceptive or illusive fulfillment of the functional requirements. And this piece of information could be useful in making a comparative assessment of automotive steering systems, for example. Regarding the longitudinal relative sliding between the rack and pinion flanks, the same characteristics are obtained in either case; they are similar to the kinematics of skiving when the pinion is made the skiving cutter to generate an internal gear, with straight or helical teeth.

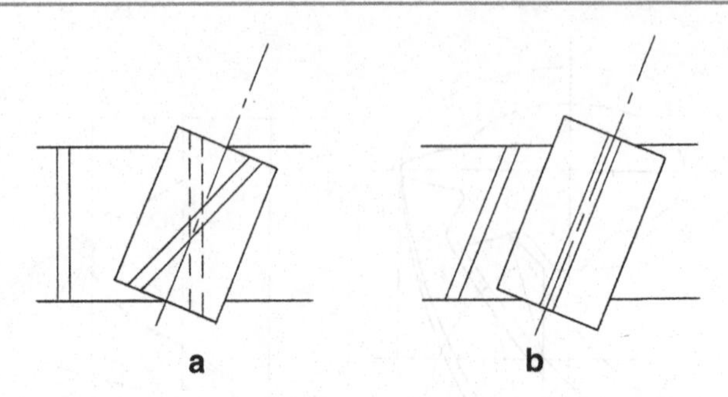

Fig. 5.11 Schematic configurations of a slanted pinion–rack pair, **a** with a helical pinion, **b** with a spur pinion

5.4.4 Ball Screw–Nut Systems

The traditional construction of ball screw and nut systems is offered by several renowned manufacturers for industrial and aerospace applications. The ball nut features one or two separate loops of ball circulation, each closed through an external return tube or an internal return passage. Two loops should provide a sort of load-sharing redundancy. One of the helicoidal ball routes is shown schematically in Fig. 5.12. The ends of the ball route are parallel, which implies that the balls leave and enter their tracks in the same spatial direction; to encompass 3½ active turns between the screw and nut threads (in this construction). In effect, the axial load is carried by eight *lower* half turns and six *upper* ones, making the design asymmetric. Since the balls that come into the upper half turn cannot—just like that —be loaded to 1.33 times the ball load in a lower half turn, a tilting moment is induced between the screw and nut. The balls will then be called upon to react this moment by changing the ratio of radial-to-axial load components; the slope of their angular-contact load path. As a result, the nut tends to tilt in the *vertical* radial plane, the upper half turns become more heavily loaded than the lower half turns, and an arduous wear pattern develops.

A more exact derivation of the tilting moment—as a function of the angle of wrap θ—is required to arrive at the *best* number of ball track turns in the nut, aiming to mitigate the effects of the inherent asymmetry of the design. Figure 5.13 shows one and a half turns of the helicoidal locus of the ball centers projected from an orientation where the ball screw axis is oblique to the plane of paper by an angle of 20° for clarity. One ball is shown at a reference position; the exit point of the ball recirculation tube and in an arbitrary, momentary position after it has traveled through an angle of wrap θ. This angle is seen correctly in the side view. The normal load F_n between the ball and the race on the screw in this latter position is resolved into the following three mutually perpendicular components, at the ball center, which lies on the pitch cylinder:

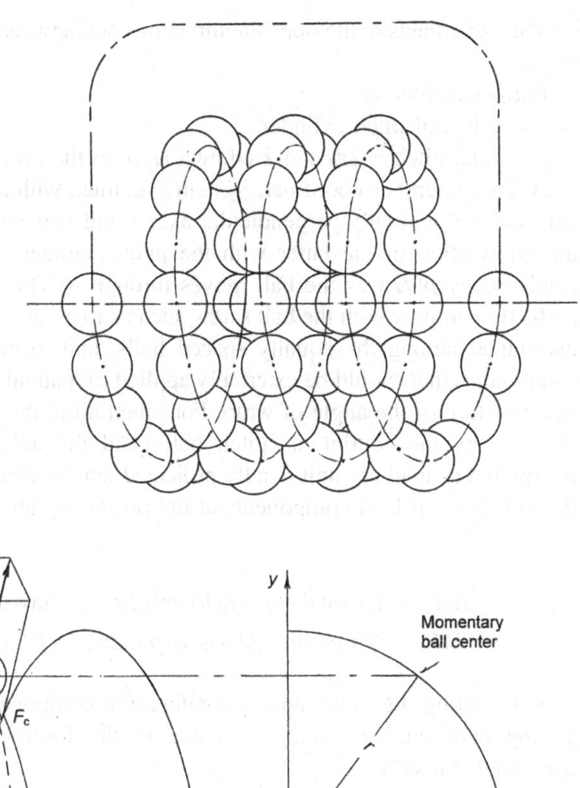

Fig. 5.12 Ball route of 3½ active turns inside a ball screw–nut system with a pitch cylinder-to-ball diameter of 5.25 and a lead angle of 5°

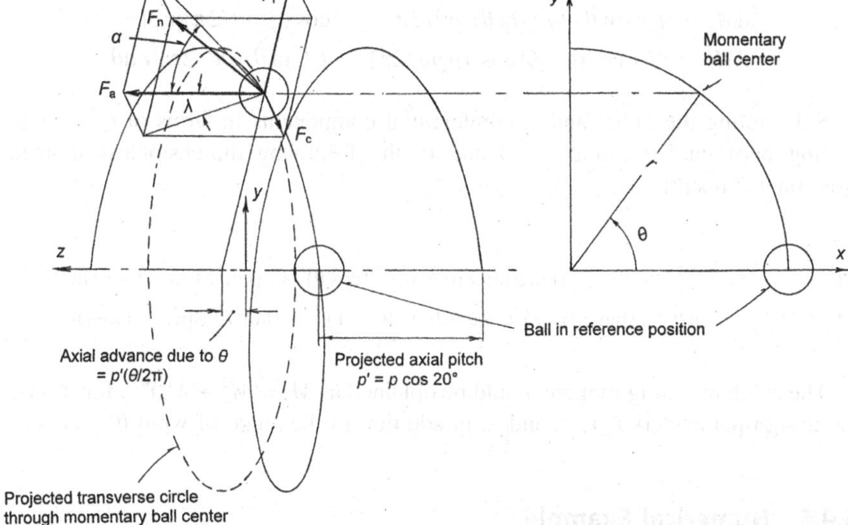

Fig. 5.13 Helicoidal ball route in a ball screw–nut system; nomenclature and force components

F_a in axial direction

F_r in radial direction; $F_r/F_a = \tan \alpha$, where α is the pressure angle between the ball and raceway in axial section

F_c in circumferential direction (tangential to the transverse circle through the momentary ball center); $F_c/F_a = \tan \lambda$, where λ is the lead angle of the helix on the pitch cylinder.

The same loads in the opposite directions act between the ball and the race inside the nut.

Further defined are:

r = radius of pitch cylinder

p = axial pitch = $2\pi r \tan \lambda$ (shown as p' in the projection in Fig. 5.13)

A fixed reference coordinate system is defined with a z-axis along the ball screw axis and two mutually perpendicular axes y and x, where the first ball center lies at the intersection of the latter with the pitch cylinder. This center will be axially displaced by $p\theta/2\pi$ as the ball moves through θ. The ball nut will be considered perfectly collinear with the ball screw such as to assume the axial load being evenly distributed among the equally spaced balls, and to calculate the tilting moment components that should be externally applied to maintain equilibrium in such a state as a function of the angle of wrap. For integration, the discrete load points will be replaced by the continuous pitch helix and the axial load component will be interpreted as load per unit axially projected length, denoted f_a. Taking moments of the *infinitesimal* load components at the position θ about the x- and y-axes,

$$\mathrm{d}M_x = [f_a r \sin\theta - f_r(\sin\theta)(p\theta/2\pi) + f_c(\cos\theta)(p\theta/2\pi)]r\,\mathrm{d}\theta$$
$$\mathrm{d}M_y = [f_a r \cos\theta - f_r(\cos\theta)(p\theta/2\pi) - f_c(\sin\theta)(p\theta/2\pi)]r\,\mathrm{d}\theta$$

Substituting the radial and circumferential components in terms of f_a and integrating between the limits $\theta = 0$ and θ, the following dimensionless moment components result:

$$M_x/(r^2 f_a) = (1-\cos\theta) - (\tan\alpha\tan\lambda)(\sin\theta - \theta\cos\theta) + (\tan^2\lambda)(\cos\theta + \theta\sin\theta - 1)$$
$$M_y/(r^2 f_a) = \sin\theta - (\tan\alpha\tan\lambda)(\cos\theta + \theta\sin\theta - 1) - (\tan^2\lambda)(\sin\theta - \theta\cos\theta)$$

The resultant tilting moment could be obtained as $M = (M_x^2 + M_y^2)^{0.5}$, function of the design parameters r, f_a, α and λ, in addition to the angle of wrap θ.

5.4.5 Numerical Example

A ball screw of nominal diameter 40 mm (r = 20 mm) and pitch p = 10 mm will have a lead angle $\lambda = 4.55$. The pressure angle is typically $\alpha = 40°$. The dimensionless tilting moment components will amount in this example to

$$M_x/(r^2 f_a) = 0.993667(1-\cos\theta) - (0.0667754 - 0.006333\,\theta)\sin\theta + 0.0667754\,\theta\cos\theta$$
$$M_y/(r^2 f_a) = (0.993667 - 0.0667754\theta)\sin\theta - (0.0667754 - 0.006333\theta)\cos\theta + 0.0667754$$

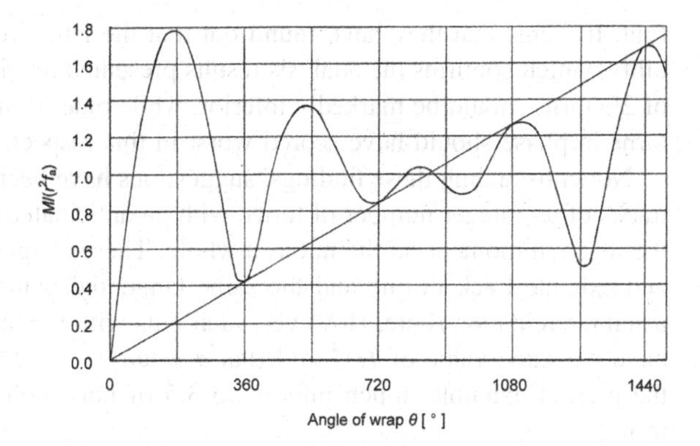

Fig. 5.14 Dimensionless tilting moment versus the angle of wrap of the balls in a ball screw–nut system

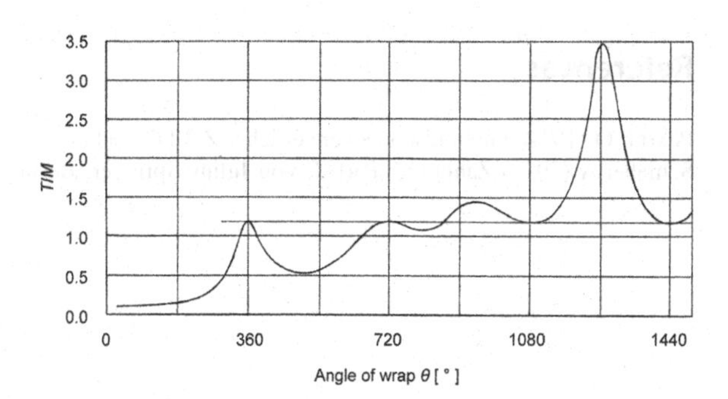

Fig. 5.15 Ratio of drive torque to tilting moment versus the angle of wrap

The resultant dimensionless tilting moment $M/(r^2f_a)$ is calculated and plotted versus the angle of wrap θ in Fig. 5.14. It is found that the tilting moment varies much as it is being integrated along the helix. It assumes a minimum value after one turn, which is of no practical value, and then again after 3½ turns.

For a more objective evaluation of adverse effects of the resultant tilting moment on the load distribution among the individual balls and on the wear pattern of the ball races, the moment should be considered in relation to the ball nut length $p\theta/2\pi = r\theta \tan \lambda$ or, better, to the drive torque $T = r^2\theta f_c = r^2 f_a \theta \tan \lambda$, both quantities expressing about the same notion, being a linear function of θ for given design parameters. According to the latter suggestion, a dimensionless figure of merit T/M is established, where

$$T/M = \theta(\tan \lambda)/\left(M/r^2 f_a\right)$$

This figure of merit reflects how small the tilting moment is in relation to the drive torque. It is calculated and plotted versus the angle of wrap θ in Fig. 5.15. The plot shows that an optimum nut ball track is one of 3½ turns, where the moment assumes a rarely encountered minimum of 1/3.5 the drive torque.

Several manufacturers offer ball screws with nuts having a track length of 2½ and 3½ turns; the odd numbers of half-turns being implied by the ball recirculation rout. It seems that they have found out that the latter version results in best durability, which confirms the analysis results presented herein. Nuts with a track length of 2½ turns should be markedly inferior, while older designs with two tracks of 1½ turns in phase should have scored worst in this respect.

Notwithstanding these findings, suggestions have been made of ball nuts to have tracks of an integer number of turns, with an anticipated objective of balancing out the reaction moment on the nut as a whole. These suggestions make the balls enter and exit the track in one and the same tangential plane, with a little longer than usual recirculation route. However, such nuts with $\theta = 2\pi$ and multiples thereof all have the same value of T/M of $1/(\tan^2\alpha + \tan^2\lambda)^{0.5}$, which amounts to 1.18643 in the present example; much inferior to 3.5 of nuts with a ball track length of 3.5 turns.

References

Pekrun O (1912) Globoidschneckengetriebe. Z-VDI 56(11):442–443
Schiebel A (1913) Zahnräder Part II. von Julius Springer, Berlin

Elastic Deformation and Microslip Issues

6

Elastic deformation of machine parts is inevitable under load and, if not properly managed, could lead to uneven stress distribution in individual parts and to unequal deformation of two jointed parts, causing them to undergo microslip, with eventual fretting hence durability reduction. The design principle of matched elastic deformation will be presented with its subordinate issues of cascades, keeping orientation, apportioned loading, and full-load bearing, all with application examples. A concise treatment of fretting, fretting corrosion, and fatigue will also be given with an eye on the design against fretting damage in machines. This is made after presenting the various mechanisms leading to microslip between the parts; not only differential elastic deformation. A real-life case study is presented on the fretting fatigue failure of a torque-free shaft end of some expensive piece of equipment. It reveals how a number of individual, unintentional design mistakes could coincide to measure up to a grave design pitfall.

6.1 Principle of Matched Elastic Deformations

Pahl and Beitz (1984) explained the principle of matched deformations by stating that "related components must be designed in such a way that, under load, they will deform in the same sense and, if possible, by the same amount." Adherence to this design principle should prevent the detrimental effects of uneven stress distribution with high-stress concentration, and of fretting corrosion due to micrometric relative displacements of the interface member surfaces under changing load. The classical example of a press fit between a shaft and a flanged hub, assuming uniform distribution of the torque along the interface, was used to explain the principle. But the proportions of the hub and shaft diameters depicted in a drawing such as Fig. 6.1 would suggest a 10:1 ratio of the polar second moments of area, making it inconvenient to talk of torsional windup of the hub to be in the same sense or opposite to that of the shaft, for deciding upon the flange to better be at the free end

H. A. Arafa, *Design for Durability and Performance Density*,
https://doi.org/10.1007/978-3-030-56816-0_6

or the inner end of the hub, respectively. For equal gross angular deformations of the shaft and hub the equal polar second moments of area would require that the hub outside diameter be 1.19 times the shaft diameter, a ratio the designer would not nearly approach. But even if this diameter ratio could be adopted and the flange was disposed at the correct end, the parabolic deformation patterns of the originally axial fibers on the shaft and hub surfaces will never coincide because they are oppositely curved.

The foregoing is but one (approximate) application example of this design principle, which could be referred to as the *principle of interface conformity*. There are a number of further important applications to cite for a comprehensive appreciation of the design principle, which are summarized in the following.

1. **Cascades** Matching piece-wise deformations of two joined parts to maintain their predetermined geometrical relationship; load to be transferred between the two parts by the cumulative contributions of a number of equal steps or stages.
2. **Keeping Orientation** Matching gross deformation components of one part to maintain the orientation of a second supported part, or of two parts to maintain the orientation of a third supported part.
3. **Apportioned Loading** Matching the deformations of two or more parts to transmit equal or proportioned loads or torques.
4. **Full-load Bearing** Shaping the deformation pattern of essentially one part to approach—under increasing load—a match with an interfaced (rigid) part in an evenly loaded contact pattern. Full matching or bearing will be achieved only under full load.

Elaborated examples of the first three applications follow. In most of the cases the design should be aided by precise finite-element analysis to arrive at an optimum configuration.

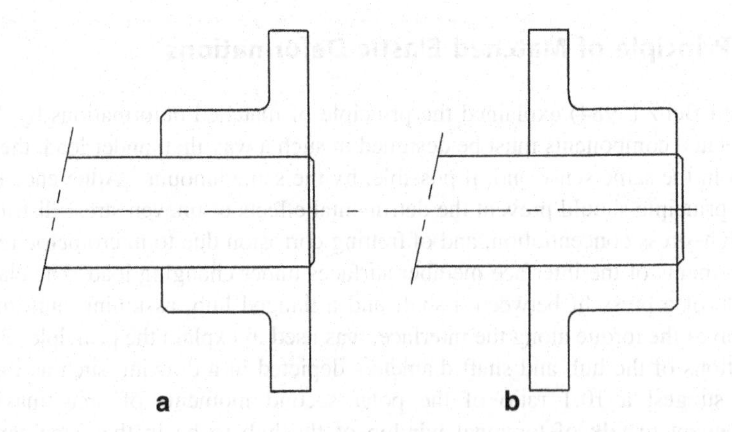

Fig. 6.1 Shaft–hub connection in a press fit, **a** hub flange at the shaft free end, **b** hub flange to the inside

6.1.1 Cascades

A typical example of matching elastic deformations of mating parts for achieving favorable load distribution at the interface as well as even stress distribution in the bulk material is the fir-tree method of turbine blade attachment to the disc or wheel of the rotor of turbomachinery; gas or steam turbines, Fig. 6.2. Each blade includes an airfoil, shank, and root sections. The blades are subjected to large centrifugal forces, in addition to bending moments, thus requiring a most effective load transfer system between the blade and the disc. Therefore, the blade root is made in form of a fir tree with a number of symmetrical pairs of load-carrying lugs to fit, very closely on the loaded side, inside a slot in the wheel rim. A small clearance is left on the distal sides of the lugs. The intermediate portions of the rim are known as wheel-posts and are made with about the same mean thickness as the blade root, to the same geometry, with a slightly less taper (due to the circular array), in an inverted sense. In lengthwise direction the fir tree could be straight or inclined, in which case the slot could be machined by broaching, or it could be curved, to be machined by form-milling. The number of pairs of load-carrying lugs could be two to five, typically three to four.

The drawing in Fig.6.2 is of a fir-tree interface of three pairs of lugs in which the fitting surfaces are all of the same inclination, in a turbine rotor of 60 blades, and it does not show the blade preloading means in the radially outward direction, to be inserted at the slot bottom. The progression of neck areas between the consecutive lug pairs is so designed as to let equal loads on the lug contact points result in equal increments of elastic elongation along the blade root and the wheel-post as well; the loads will thus remain equal at all points. The blade will then be pulled by the cumulative contributions of a number of nearly equal co-acting forces, which configuration is often called a cascade.

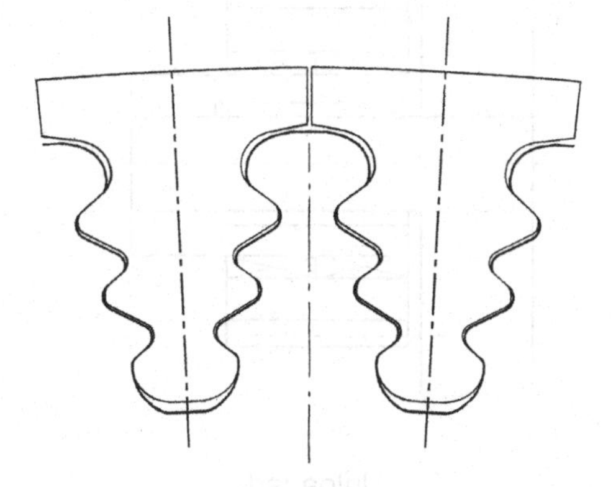

Fig. 6.2 Fir-tree attachment of turbine blade roots to a rotor disc

6.1.2 Keeping Orientation

In planetary gearing single-sided carriers allow a maximum possible number of cantilever-supported planets to be used, with minimum tooth tip-to-tip clearance. This makes the planetary set of highest possible torque density.

To achieve equal load sharing by the planets on a single-sided carrier, Hicks (1967) suggested that the bearings of each planet be supported on a cylindrical sleeve of substantial thickness, which is then cantilever-supported on the end of a cantilever flexible pin such that the resultant of the two tangential tooth loads (acting in the planet midplane) lies in the middle of the free length of the flexible pin. This load will thereby exert equal and opposite bending moments on the two fitted ends of the pin, which will then remain parallel as the pin is transversely deflected. This principle is shown in Fig. 6.3. The deflection under rated load is made markedly greater than the expected manufacturing errors which would otherwise cause uneven load distribution among the planets. This design also makes the planet virtually self-aligning, tending to maintain full-face contact with the sun and/or the ring.

Both the original patent by Hicks (1967) and the paper by Hicks et al. (2004) depict a plain cylindrical pin that should be heavily press fitted both in the planet carrier and the sleeve—may be only for a schematic demonstration of the principle. The two edges of contact in such an assembly are at the maximum bending curvature of the pin and would thus be prone to fretting. Contouring the pin to provide

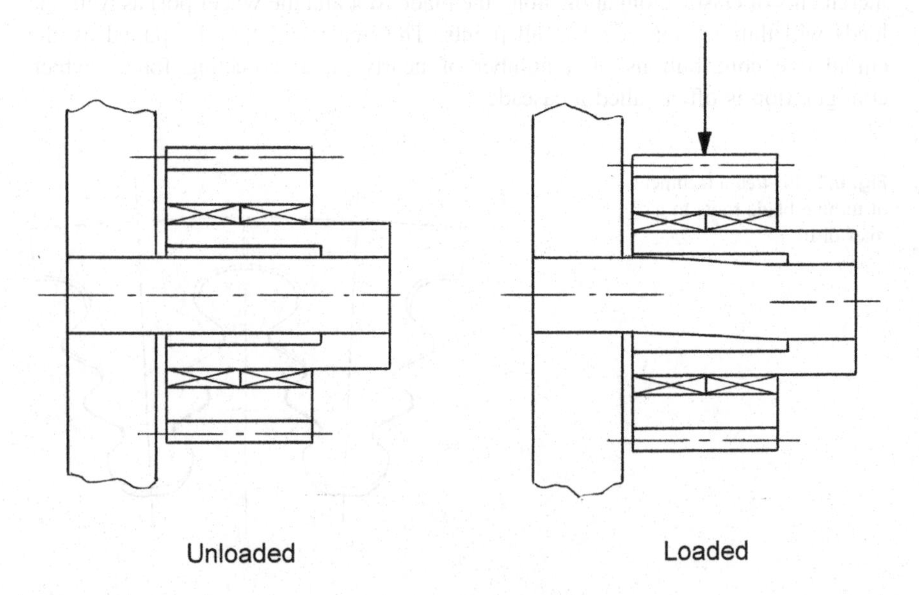

Fig. 6.3 Flexible pin support of a planet gear on a single-sided carrier

larger-diameter fits as well as alternative means of securing the sleeve to the flexible pin for easier assembly and replacement of the parts are suggested for example by Fox (2014).

6.1.3 Apportioned Loading

Example 1: Multi-part Spur Pinions
Power transmission between a slender pinion and a larger gear is known to suffer substantial pinion windup, so that edge loading ensues. This effect could be counteracted by dividing the pinion transversely into two (or three) parts and connecting them to the drive end through equal-compliance quill shafts. The pinion parts will then equally share the torque and their shafts equally twist. Figure 6.4 depicts a two-part pinion with two quill shafts, where the outer pinion is supported on the inner one, which is then supported on an outboard bearing. The pinion blanks should first be assembled through a closely fitting spline connection then the pinion teeth simultaneously cut and finish-ground.

Example 2: Dual Spiral Bevel Gearing
A similar idea has been suggested by Fischer et al. (2003) for the spiral-bevel-gear final reduction stage of rotorcraft transmissions. Bevel and spiral bevel gears are customarily designed with only limited face width; the tooth toes would otherwise become too small. For achieving higher power density a dual power path could be used, where two concentric spiral bevel wheels (as if in one-piece) are driven by two coaxial spiral bevel pinions that are connected to the input drive by two quill shafts. The two gear sets should be of the same reduction ratio, which—should that ratio not be an integer—is only possible by having the gears of the same number of teeth, so as the pinions, and the modules of the two sets in the same ratio as their sizes. The two quill shafts should be of such proportioned compliance as to transmit correctly apportioned torques under equal windups; the smaller of the two gear sets transmitting a smaller torque. However, the phasing problem could not be handled by such techniques as simultaneous cutting and/or finish grinding, such as with the afore-mentioned multi-part spur pinion, and the assembly is further complicated by the need for fine axial adjustment of the two pinions individually for achieving good tooth-bearing patterns. It does not seem that such a design has been approved for production.

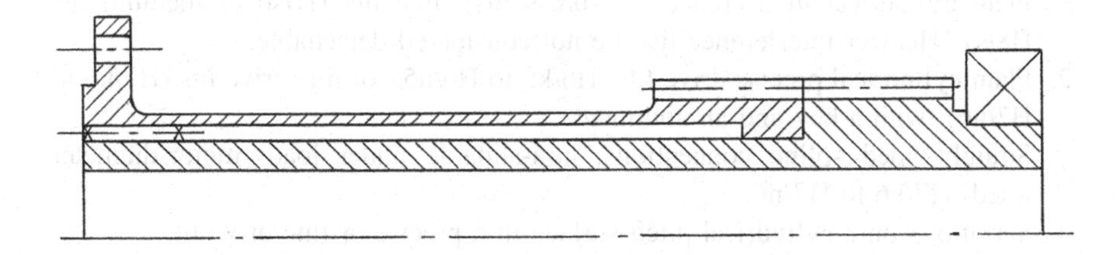

Fig. 6.4 Two-part pinion for equal load sharing

6.1.4 Full-Load Bearing

Example 1: Slender Pinion Windup Compensation

Slender pinions should be finish-ground with a very small helix angle (increment) in a sense as to nullify as close as possible their torsional windup in operation. This is a well-established practice in gear manufacture. Complete matching or full-face contact will be achieved only under full load, and in only one loading direction.

Example 2: Splined Shaft–Hub Joints

Splined shaft–hub joints suffer torsional windup that makes the splines bear much of the load at the torqued shaft end, particularly when the spline length is substantial. This could be overcome by curving the male splines according to a slightly parabolic function in axial direction; the female splines being broached as usual. A little larger clearance should be left for assembly.

6.2 Jointing Detachable Static Interfaces

Jointing detachable static interfaces is a subassembly step within the process of assembling mechanical equipment. It is done for some out of several reasons:

1. Bringing standard bought-out components—usually of different technology—into the assembly. Examples: rolling bearings, multi-disc clutches, and brakes.
2. Integrating specialized components of higher technology and of different material into the assembly in so-called repetitive modularity. Examples: blades of axial-flow turbomachinery.
3. Re-jointing split or partitioned components for accessibility to assemble with other parts. Examples: split big-ends of engine connecting rods, split-race rolling bearings, block-type universal joints.
4. Making a component in two or more pieces for manufacturability; saving on the material to be cut away; bulkiness build-down. Examples: shaft–hub connections, built-up engine crankshafts, large rim-gears.

The *direct* jointing interface could assume any of the forms listed below, to be provided with fastening or retention means when needed. Jointing through intermediary expanding elements is not considered in the present context.

1. Plain cylindrical interference fits (press fits); fine fits H7/s6 or medium fits H8/s7. Heavier interference fits are not considered detachable.
2. Plain cylindrical precise drive fits (H6/k5 to H6/n5), or fine drive fits (H7/k6 to H7/n6), with a key and an abutment.
3. Straight-sided spline connections, non-slidable under load, minor-diameter fitted, H7/k6 to H7/n6.
4. Serrations on a cylindrical pitch surface in a precise or fine drive fit.

5. Flat with two or more locating pins.
6. Flat with parallel serrations and one or two locating pins.
7. Flat with two sets of parallel serrations perpendicular to each other.
8. Tapered, self-holding; 1:8–1:16.
9. Slightly tapered, 1:20–1:30.
10. Face serrations (Hirth couplings).
11. Face-toothed couplings.
12. Blade roots; dovetail, fir tree, and similar types.

The jointed detachable static interface should fulfill the following functional requirements:

1. Allowing (easy) assembly and disassembly.
2. Location with repeatable accuracy.
3. Load transfer by positive or frictional means, as envisaged, without suffering microslip at any surface patch under full load.

6.3 Fretting Damage in Machines

Fretting damage remains one of the plagues for mechanical systems; it is often the root cause of loss of functionality or catastrophic failure. Although fretting damage has been reported and investigated since the late 1930s it is still one of the most difficult problems to face. Essential outlines of the problem and only samples of design solutions to it are given hereinafter.

Fretting Fretting is the cyclic microslip—reversed or unidirectional—between two metallic contacting surfaces, when sustained under normal load. There are two main cases of fretting surfaces and their associated micro-motion:

1. Conformal contacting surfaces with oscillatory tangential microslip in either a circumferential or an axial direction, with an amplitude <50 μm, or in slow continuous circumferential shift or creeping in form of a slip wave, due to traveling normal point loads.
2. Non-conformal surfaces in line-contact under oscillatory tangential microslip, usually normal to the contact line, or also under cyclic radial loads or cyclic twisting or drilling micro-motion.

Lubrication could help reduce fretting, but it can also accelerate it if the lubricant allows more of the microslip.

Fretting Corrosion This is a combined action of fretting and corrosion. Therefore, the presence of oxygen is important; being in the air or dissolved in the lubricant. Microslip under load continually welds together then strips off the contacting asperities on the mating surfaces. The resulting fine wear particles (debris) as well as the exposed, active metal surfaces oxidize, to become much harder than the original metal. With line-contact fretting the debris will not be retained where it is produced, but with surface-contact fretting it will remain trapped inside the interface, where it acts as an abrasive to further aggravate the fretting wear. The oxide layers are repeatedly formed and rubbed away by the abrasive action, leading to scoring and surface pitting. The debris is reddish-brown on iron and steel, and black on aluminum and its alloys.

Fretting Fatigue Fretting corrosion causes surface disintegration due to wear and the subsequent formation of shallow pits, filled and surrounded by the debris, but it is the detrimental effect on the fatigue endurance of the components that is of the most concern. Remember that 90% of the fatigue life goes into just starting a crack. The pits caused by fretting will act as stress concentrators. Fatigue micro-cracks are initiated close to the boundary of such pits; they generally start obliquely to the surface and propagate through the bulk material to failure. Fretting fatigue is one of the important phenomena that cause unexpected failure accidents of mechanical equipment; cracks can initiate at very low-stress levels, well below the fatigue limit of non-fretted parts.

6.3.1 Microslip in Loaded Conformal Contacts

The occurrence of microslip in loaded conformal contacts could be attributed to any one of the following sources/reasons.

1. Differential elastic deformation, usually in cylindrical shaft–hub connections in torque-transmitting fits, particularly when the torque loading is at the improper locations (the same-side ends) and the fit is insufficiently tight, in non-compliance with the interface-conformity mode of the design principle of matched elastic deformations. The normal load in the interface is the radial stress due to the fit, and microslip starts at the *edge of contact* on the loaded end. Fretting then occurs when the assembly is under variable or cyclic torque loading.
2. Kinematics of keyed, close-running cylindrical fits (not intended to transmit any appreciable torque) where a hub is insufficiently clamped against a shaft shoulder and a stationary tilting moment acts on the rotating members to produce cyclic rocking or jolting. The larger the abutment radial extension above the shaft the larger will be the resulting cyclic axial microslip at any given point in the shaft–hub interface. The location of the spots first exposed to fretting depends on the form errors of the two surfaces in contact.
3. Kinematics of close-running and push cylindrical fits, keyed near one end, where rotating rocking or jolting due to a stationary tilting moment on the rotating members produces orbiting relative motion, within the clearance, at the other end. The rotating rocking or jolting moment should be sufficient to produce a sensible normal load at the interface. The orbiting motion produces circumferential microslip of a repetitive nature on any given spot, rather than a cyclic one.

4. Interface creep due to traveling wave deformation; the flexing micro-movements of a rather slim member under moving point load(s) over a relatively rigid subsurface. The parts could be straight or, more usually, round such as radial rolling bearing rings. In the latter case the bearing would be the non-locating one, with a light fit on the ring. These movements occur without nominal torsional loading, and they create a unidirectional slip wave in the bearing seat, leading to a continuous shift, hence fretting. One solution is to use the (slightly) toroidal roller bearings, which are known to accommodate some angular misalignment as well as axial freedom, with virtually no increase in friction. Using such bearings obviates the need to compromise between a tight fit and axial mobility, permitting tighter fits to be adopted to eliminate creep and fretting corrosion.
5. Loaded interfaces established with parts that are additionally included in a machine just to eliminate kinematic overconstraints (see Sect. 7.2). These interfaces will undergo microslip that increases in magnitude as the accuracy of manufacturing the machine components worsens.

6.3.2 Overconstraint-Relieving Joints Prone to Fretting

Examples are taken from suggestions for engine mechanisms to be free of kinematic overconstraints (see Sect. 7.3).

Example 1 The conventional connecting rod is replaced by a three-piece articulated one as shown in Fig. 6.5a. The revolute joints between the piston pin

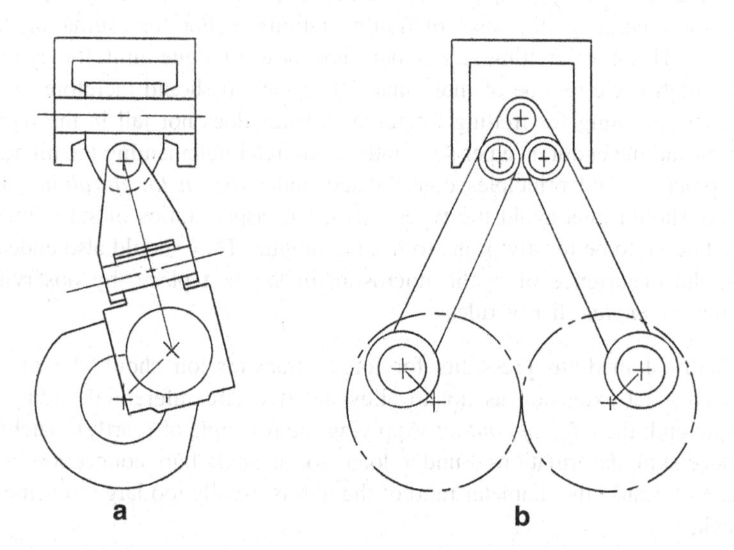

Fig. 6.5 Engine mechanisms with unconventional connecting rods, **a** three-piece connecting rod, **b** two connecting rods with a triangular balance link, for two crankshafts in anti-phase

and the connecting-rod first block and between the connecting-rod middle part and its big-end bearing block do not actually rotate. They are there only to remove the two overconstraints typical of conventional engine mechanisms (in spatial perception); one alignment and one lateral overconstraint (see Sect. 7.3.4). These joints will be prone to fretting. However, no engine manufacturer would adopt this construction because the lateral overconstraint between the connecting-rod small end and the piston is removed by allowing a small axial clearance between them, and the alignment overconstraint between the connecting-rod big end and the crank is readily taken by the oil film thickness of the hydrodynamic bearing between them, in view of the high accuracy of manufacture of the engine parts.

Example 2 Consider an engine mechanism with two connecting rods per piston (having two piston pins) and two counter-rotating, geared crankshafts, for balanced side thrust. The idea is limited to short-stroke engines to avoid collision between the connecting rods and the cylinder. A piston with one pin could again be used and a triangular balance link added between the piston pin and the two small ends of the connecting rods, as shown in Fig. 6.5b, to remove the one overconstraint (in a planar perception of the mechanism). This leaves the piston to be freely guided in the cylinder, unaffected by phasing mismatch of the two gears or any other dimensional inaccuracies of the components. The revolute joint of the triangular link with the piston pin does not actually rotate; it will be prone to fretting.

6.3.3 Designing Against Fretting Fatigue

Some scientific researchers call the designing against fretting fatigue "the least quantitative of all fatigue topics." This may be due to them mainly searching for models for predicting the onset of fretting fatigue and/or for estimating fretting fatigue life. However, fretting fatigue does not show a fatigue limit; it may occur in the very high cycle regime of more than 10^7 cycles. It should therefore be appreciated that designing for fretting fatigue avoidance does not fall in the regime of design for fatigue endurance per se; it rather requires implementing the higher-level design practices and principles consolidated under *design for durability*. Design engineers should thus avoid the types, sizes, and aspect ratios of static interfaces that are known to be mostly prone to fretting fatigue. They should also endeavor to prevent the occurrence of cyclic microslip, in the first place, by observing the following *incomplete* list of rules.

1. Cylindrical medium press fits for torque transmission should be of such a limited axial extension as not to allow relative circumferential sliding at the torque end; the *edge of contact*. Applying the principle of (partly) matching the surface/skin deformations—under load—of a shaft–hub connection is often useless because the diameter ratio of the hub is usually too large to achieve this match.

2. Cylindrical interference fits for supporting cantilever loads should be avoided, or else be of sufficient diameter and axial extension to prevent relative axial microslip.
3. Opting for the components that accommodate misalignment by elastic deformation in lieu of those that undergo relative sliding displacements; flexible-frame couplings instead of gear couplings, for example.
4. Providing sufficient clamping load on flat and wedge-action (serrated) detachable static interfaces to mitigate fretting fatigue.
5. Abiding by recommended, most appropriate fits between rolling bearing races and their shaft and housing surfaces to prevent slip-wave creeping.

6.3.4 Example of Designing Against Fretting Fatigue

Supporting a pair of planets on both sides of a bogie plate (planet carrier) could be achieved by press fitting the planet pin in the bogie plate, in a double-cantilever fashion, such as in Fig. 6.6a. Minute imperfections in the gearing dimensions and geometry would lead to the press fit tending to rock a little, at random, which subjects it to fretting corrosion. The fit tightness could thus be lost, which could result in catastrophic failure of the system. This can be mitigated when the pin halves are machined back-to-back in one-piece with a slim, elastic part of the bogie plate, such as in Fig. 6.6b. The latter suggestion is after Gravina (2017), but it requires sophisticated manufacturing processes.

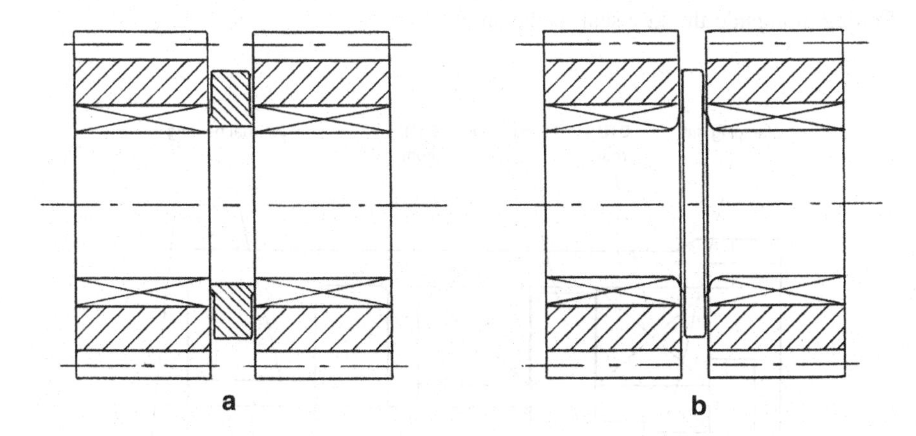

Fig. 6.6 A pair of planets supported on both sides of a planet carrier, **a** on a separate double-cantilever planet pin, **b** on two planet pins integral with the carrier

6.4 Case Study

Fretting fatigue failure of a torque-free shaft end

System under Investigation Horizontal multi-stage centrifugal pump, shaft outboard end supported in a journal bearing and axially restrained by a double-acting, equalizing tilting-pad thrust bearing (ETB) on a separate thrust collar to react the residual axial load on the shaft as shown in Fig. 6.7. The residual axial load is the inevitable difference between the hydrodynamic thrust of the impellers and the opposite hydrostatic thrust of the balance drum (piston), whichever dominates to define the direction of the axial load and which of the thrust bearings should take it. The impellers and the balance drum are not shown in the drawing. Bearing size is 9 inch, pump rated power ≈2,000 kW at 2,900 rpm, driven by an electric motor by a coupling at the drive end.

Reported Incident The shaft outboard end broke inside its thrust collar hub after about 7000 working hours. This happening was considered a *weird encounter*, since that end is almost torque-free and no substantial dynamic load in that region is expected, either.

6.4.1 Shaft Failure Characterization

Visual Inspection Result
A complex fretting corrosion pattern with areas of reddish-brown coloration of varying intensity, associated with pits, was obvious all round on the cylindrical shaft–collar interface. In particular, this pattern included two strips of rather heavy fretting alongside the key seat in the shaft.

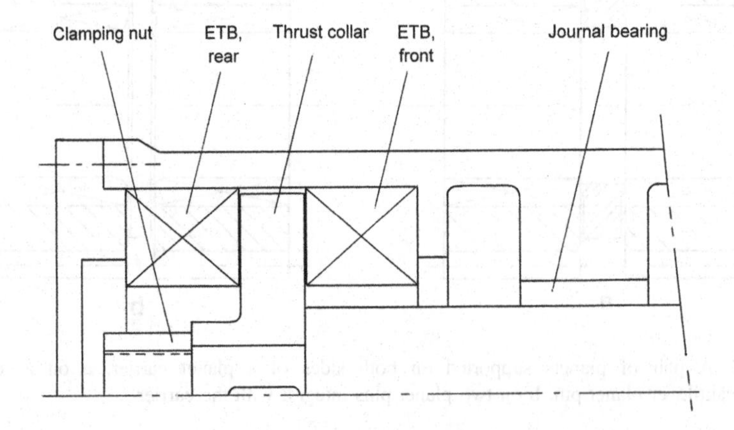

Fig. 6.7 Schematic of the bearing system of the shaft outboard end of a horizontal multi-stage centrifugal pump, shown without the lubricant drain sump

Closer Inspection Result

Scanning electron micrographs of the fretted strips adjacent to the two straight key seat edges showed surface damage due to microslip in the axial direction. The fractured shaft section showed that a fatigue crack had originated at the edge of the key seat in the heavily fretted area, at about the collar hub mid-length, rather than at a typical geometric stress raiser such as the key seat bottom corners. The fracture surface exhibited clear beach marks, characteristic of fatigue failure, which were generated as the fracture propagated across the section. The smooth fracture area covers about 97% of the shaft cross-section, and the remaining 3% exhibits a rough ductile fracture. These features indicate that the main cyclic load causing fatigue failure is rotating bending of a small magnitude, compared to the load-carrying capacity of the shaft.

The Thrust Collar

Fretting (alone) is a known failure mode in the shaft–collar interface of ETBs, mostly witnessed by those working with rotating machinery in the petroleum business. To prevent fretting, the thrust collar should be either integral with the pump shaft or a separate part, in which latter case it should be mounted by hydraulic shrink fitting on the pump shaft; as an acceptable and reliable design. However, many designs still incorporate a separate thrust collar on a cylindrical fit with the shaft, such as in the present case study. Therefore, a design appraisal was conducted to bring forward the features and practices adopted in the pump design, which could have an effect on its shaft tail-end endurance; service to failure in fretting fatigue. Most striking of all was that, for some reason, the pump designer decided to depart considerably from the bearing manufacturer's recommendations regarding the separate thrust collar and shaft end dimensions, which are shown in Fig. 6.8a, to adopt the design shown in Fig. 6.8b, both drawings being to the same scale, for the same bearing size.

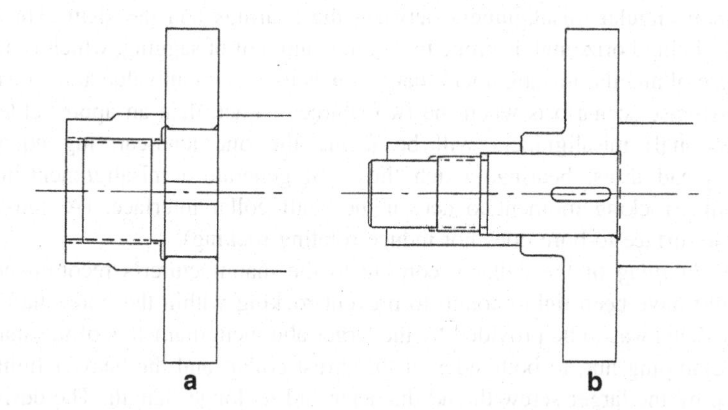

Fig. 6.8 Proportions of a separate thrust collar of an equalizing tilting-pad thrust bearing, **a** according to the manufacturer's recommendation, **b** in the Case Study 6.4

The relevant differences between the two designs, viz., the anticipated *savings* are summarized as follows.

1. The thrust collar bore was reduced from 88.9 mm (3.5 inch) to 57.15 mm (2.25 inch).
2. The shaft shoulder (abutment) diameter behind the thrust collar was reduced from 108 to 100 mm.
3. The clamping nut diameter retaining the thrust collar was reduced from 108 mm to 70 mm.
4. The screw thread nominal diameter was reduced from 73 mm (2 7/8 inch) to 51 mm (2 inch).

6.4.2 Root Cause Analysis

1. The two heavily fretted strips alongside the key seat indicate that the shaft surface was a little elevated around the key seat. This feature could have its reason in a process of aggressively machining the key seat that plastically deforms its environment or, more probably, due to wrongly *heavy press fitting* the key.
2. Microslip in the shaft lengthwise direction can only be due to cyclic rocking or jolting of the thrust collar under misalignment-induced rotating bending moment in the equalizing tilting-pad thrust bearing (see Sect. 4.2). The larger the abutment shoulder behind the thrust collar the larger will be the amplitude of the axial micro-motion at any given point on the shaft periphery.
3. The bearing housing is a separate, axially split component bolted onto the end face of the pump casing. Due to manufacturing tolerances it could be a source of a small angular misalignment between the bearings and the shaft. The pump shaft, being horizontal, is prone to a minute amount of sagging, which is another source of angular misalignment that assumes its maximum value at the bearings. Worst-case scenario is when the two sources co-act, then an appreciable (unidirectional) misalignment will be facing the one load-carrying equalizing tilting-pad thrust bearing, which thus will generate a misalignment-induced rotating rocking moment to act on the shaft–collar interface. (A non-square collar surface-to-bore does not induce rotating rocking).
4. The *clamping* of the collar according to the manufacturer's recommendation should have been tight enough to prevent rocking within the bore–shaft clearance. This was to be provided by the larger abutment diameters of the shaft and the clamping nut, to both sides of the thrust collar, and the heavier tightening force by the larger screw thread diameter and its longer length. The designer's decision to trade these features for largely inferior ones obviously allowed the thrust collar hub to rock relative to the shaft (at the frequency of its rotation),

which led to fretting. The heavily press-fitted key intensified the fretting process around the key seat edge, which caused the subsequent fretting fatigue rupture of the too thin shaft. This design decision was obviously made in ignorance of the effect of the pivotal stiffness of the thrust bearing that produces a rotating bending moment upon the slightest angular misalignment.

Experience Gained from this Case Study

A number of individual, unintentional mistakes could coincide to reveal the consequences of a grave design pitfall that leads to unexpected failure of an expensive piece of equipment.

References

Fischer M, Hunold B, Häse H et al (2003) Gearbox with torque division, in particular for a helicopter rotor drive. US Patent 6,626,059, 30 September 2003

Fox GP (2014) Epicyclic gear system with semi-integrated flexpin assemblies. US Patent 8,920,284, 30 December 2014

Gravina M (2017) Planet-carrier for an epicyclic gearing and epicyclic gearing provided with such planet-carrier. US Patent 9,702,451, 11 July 2017

Hicks RJ (1967) Load equalizing means for planetary pinions. US Patent 3,303,713, 14 February 1967

Hicks RJ, Cunliffe F, Giger U (2004) Optimised gearbox design for modern wind turbines. 2004 European Wind Energy conference (EWEC), London

Pahl G, Beitz W (1984) Engineering Design. The Design Council, London and Springer, Berlin

7 Assembly Over Mobile Interfaces

The kinematic properties of mechanical assemblies regarding their two aspects of the mobility and the state of constrainedness will be addressed. Constraint analysis of such assemblies is first based on the assumptions of infinitely rigid bodies and perfect geometry of the interfaces or joints, taking the anticipated mobility and any idle or independent mobilities into consideration. The resulting count of kinematic overconstraints, if any, turns out to be a very useful marker, to be taken in combination with the accuracy of manufacture of the so-called inter-joint bodies of the components and their inevitable elastic deformation under load, for making a critical assessment of any proposed design regarding assembly, functionality, and durability. Examples are presented of quasi-exactly constrained devices and how they represent well-proven designs, and of overconstrained devices that could result in catastrophic premature failure of expensive pieces of equipment.

7.1 Assembly

Assembly in mechanical engineering is the putting together, retaining, securing, or else tightening of all the parts and components required to construct the mechanical system in definite accordance with its design. In doing so, there emerge static/immobile interfaces as well as mobile interfaces.

A single-part component will not have any function all by itself, even parts that seem to function on their own, springs for example, do not. Rigid mechanical parts acquire functional characteristics only when assembled or mated to each other over mobile interfaces. Taking any one part as a fixed reference, another part will, before assembly, have six degrees of freedom (DOF) of motion relative to the reference part. Interfacing the two parts is an action that reduces the number of relative DOF to less than six, by imposing a number of constraints from one to five. This number is also referred to as the connectivity of the interface. An ideal single constraint is obtained through a single-point contact. Therefore, a cubic part will be exactly

H. A. Arafa, *Design for Durability and Performance Density*,
https://doi.org/10.1007/978-3-030-56816-0_7

constrained by making it rest over one, two, and three single points on each one of three mutually perpendicular faces. This condition can only be acceptable when loads are relatively small, for example in instrument design. For the interfaces to transmit load, the three-point contacts are replaced by a flat surface and the two-point contacts are replaced by a straight-line contact.

A compilation of the basic single kinematic interfaces in 3-D perception is presented in Fig. 7.1, classified into the only six conformal surface contacts and examples of straight-line and circular-arc contacts, each with its connectivity. Out of these cases, only the revolute and the plunging (prismatic) pair could be present in planar assemblies as well. This compilation is an essential guide in performing constraint/mobility analysis of a mechanical assembly. Some complementary definitions are due in the following.

Conformal Surface Contact The single contact between continuous surfaces of two rigid bodies, allowing only relative sliding.

Straight-line Contact The single contact between two rigid bodies on a straight-line that allows relative sliding along and perpendicular to the line as well as relative angular motion (and any combination thereof). Should the two bodies be set in purely angular relative motion, the straight-line of contact must not be the instantaneous axis thereof.

Circular-arc Contact The single contact between two rigid bodies on a (planar) circular-arc which allows relative sliding along the arc and perpendicular to its plane, in such settings that should also allow relative angular motion of the two bodies; in the sense of rolling or also of angular out-of-alignment from a reference position. Circular-arc contacts that do not allow such (tube edge on a flat surface for example) are considered conformal contacts, rather.

Specifying a kinematic pair or mobile interface in a mechanical assembly of two rigid parts implies for both their surfaces that are intended to come into contact, conformal or non-conformal, the following:

1. Continuity or pseudo continuity of both surfaces.
2. Geometric exactness: flatness, roundness, cylindricity, straightness, taper, sphericity, and ovality or lobedness (of extruded features).
3. Equality of the key dimensions: single-parameter specifications such as the radius of cylinders and spheres and the cone angle, or two-parameter specifications such as the base cylinder radius and pitch of involute helicoids, the major and minor axes of ellipses, and the ring and tube radii of tori.

7.2 Mobility of Mechanical Assemblies

A mobile mechanical assembly consists of a number of rigid components assembled together over mobile interfaces or joints, either in series, parallel, or a combination thereof. The mobility (abbreviated M) or the number of degrees of freedom of a mechanical assembly is defined as the number of mutually independent

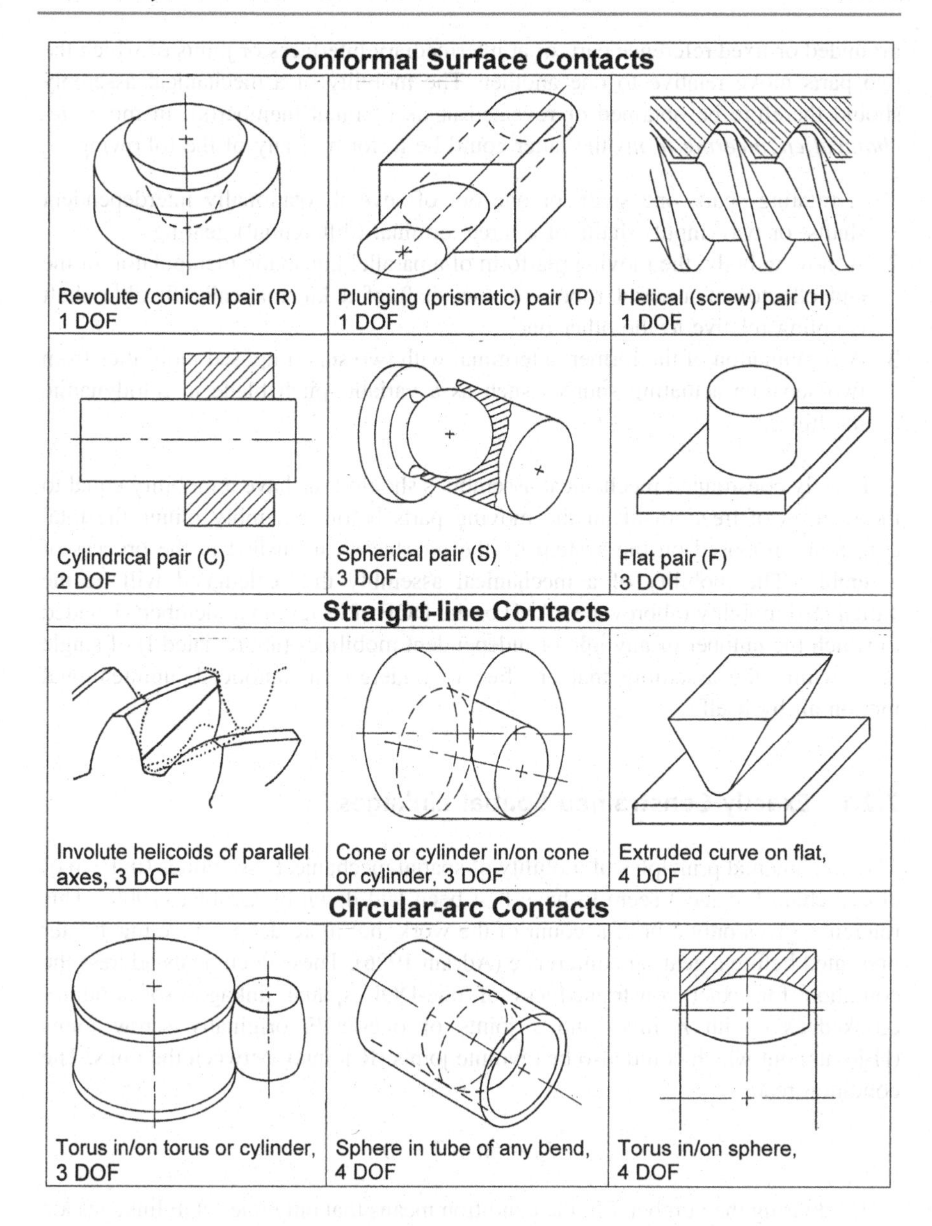

Fig. 7.1 The basic single kinematic interfaces

position variables necessary and sufficient to describe the state of the assembly in terms of the mutually orthogonal position components of all of its parts. (The latter may be a much larger number than the assembly mobility itself). The position variables could be measured or introduced at parts that move relative to the grounded or fixed reference part, or at intermediary interfaces or joints of which the two parts move relative to one another. The mobility of a mechanical assembly should therefore be obtained or revealed at its terminal member(s). In more *mechanical engineering terms* the latter could be in form of any of the following.

1. A rotating shaft; one shaft, or one out of several rotationally interdependent shafts, or two (input) shafts of a three-terminal (differential) gearing.
2. A moving body; the moving platform of a parallel kinematic manipulator, or the end effector of a serial robot, or one shaft of a kinematically flexible shaft coupling relative to the other one.
3. A combination of the former; a terminal with two superimposed mobilities from two separate actuating sources such as a variable-pitch blade in rotodynamic machines.

Exactly constrained mechanical assemblies should thus have a mobility equal to the degrees of freedom of all the moving parts before assembly minus the total constraints imposed on the system by the interfaces established in the process of assembly. The mobility of a mechanical assembly thus calculated will be the anticipated mobility (abbreviated AM) or the DOF of its terminal member(s), added to which the number of any idle or independent mobilities (abbreviated I) of single parts within the assembly that are free to undergo an unnoticed, unintentional motion all by itself.

7.2.1 Exactly Constrained Spatial Linkages

The fundamental principles of mobility of spatial mechanical assemblies (in form of closed-chain linkages) seem to have first been laid down by Grübler (1909). This reference gives only a brief account of the work, but more details were much later presented before a gearing conference (Adrian 1926). These documents address the condition for exact constrainedness of one-DOF spatial linkages (of arbitrary crookedness) with n links and s joints of one-DOF, originally screw joints (H-joints) but which could also be revolute joints (R-joints), between the links. The condition reads

$$5s - 6n + 7 = 0$$

Hardwiring the number 7 in the condition means that only one-DOF linkages are treated;

$$1 = 6(n - 1) - 5s$$

with the corollary that only a seven-bar chain with seven joints of one-DOF each would be exactly constrained to have one-DOF. However, this expression is valid for any exactly constrained, spatial multi-DOF linkage having one-DOF joints, with a mobility M instead of the (1) on the L.H. side;

$$\mathrm{M} = 6(n-1) - 5s$$

Grübler's deliberations were confined to exactly constrained one-DOF linkages, may be by reason of

- The accuracy of manufacture at that time obviously did not allow dealing with overconstrained mechanisms; these would either not assemble or just not function; jam.
- Multi-DOF mechanisms were not that interesting as they are now for several applications such as in manipulators.

Obviously due to the cumbersome appearance of seven-bar linkages with screw joints, mention was made in Adrian (1926) of reducing the number of links in steps of removing one and connecting the adjacent links through a cylindrical-joint (C–joint), and/or removing two consecutive links and connecting the adjacent ones by a spherical joint (S–joint). This would result in four-bar C–C–C–H(R) and three-bar C–S–C linkages, respectively.

7.3 Kinematic Overconstraints

In 3-D mobile assemblies, multiplying the number of moving parts by six and subtracting the number of constraints imposed by the jointed interfaces produces the so-called Grübler's count (abbreviated G), rather than the mobility proper. In some cases, this number may represent the anticipated mobility of the system (including any idle mobilities). In these cases, the mechanism is properly or exactly constrained. The advantageous features of an exactly constrained assembly are that, regardless of the accuracy of manufacture, and although all the parts are assumed rigid.

1. The parts will assemble in any position throughout the range of operation.
2. The device will smoothly execute the full range of operation without a tendency to jam.
3. The assembly will be statically determinate and will keep all the interfaces evenly seated even if the bulk material would deform under load.

This is because the shapes or geometry of the inter-joint (or inter-interface) bodies of the members, however *crooked*, are irrelevant to the kinematics; the geometric relationship between any two features in one and the same rigid body is not taken into account in the constraint analysis. However, in an exactly constrained, crooked mechanism the kinematic relationship between the input(s) and output(s) will be distorted from the theoretical, according to the degree of crookedness.

In several cases Grübler's count will however be smaller than (AM + I), often going into negative numbers, which fact has tempted some teachers to wrongly convey to their mechanical engineering students the notion of "paradox." This is because the counts of AM and I are rather easily determined by inspection, and it would not be awaited until G is calculated to know of the mobility of the device. In such cases of discrepancy the assembly should be identified as being *overconstrained*, by virtue of the constraints imposed by the interfaces being more than just appropriate for exact constrainedness; the difference between the two results being the number of kinematic overconstraints (abbreviated O) concealed in the assembly. Therefore,

$$AM + I = G + O$$

This AMIGO expression relates the kinematic properties of assemblies regarding mobility and the state of constrainedness; it is used for *constraint analysis* of an assembly. It must be emphasized that kinematic overconstraints and idle mobilities are not mutually exclusive; one assembly could feature both aspects simultaneously. Should I and O be equal, then the anticipated mobility will equal G, incidentally. It should also be emphasized that an open chain of serial members cannot be overconstrained.

It turns out that calculating G, hence O, is a very useful marker for assessing a given mechanical assembly of rigid parts as being exactly constrained or overconstrained, rather than for determining the mobility itself. And it should be remarked that the scientific community had witnessed numerous efforts just for determining the mobility. For example, Gogu (2005) made a critical review of calculating the mobility according to thirty-five approaches/formulas of the last 150 years. He concluded that these formulas do not fit for many "classical or modern" mechanisms. This conclusion could have been anticipated, since the paper is devoid of the word *overconstraint*.

Kinematic overconstraints may be classified into the following types:

1. Plurality overconstraints.
2. Angular overconstraints.
3. Lateral (or linear) overconstraints.
4. Phasing overconstraints.
5. Intrinsic overconstraints.

Combinations of different types of overconstraints in a mechanical assembly are possible.

7.3.1 Plurality Overconstraints

Multi-Coupler Drive
A multi-coupler drive is shown in Fig. 7.2a; it is one type of the so-called close-center drives, used to drive one shaft from another at a small, fixed parallel offset. It consists of two discs that are interconnected by a set of three (or more) parallel couplers, of such a length as not to collide in operation. Without the shafts and bearings, as shown, the drive has four moving parts, hence it conceals one plurality overconstraint; one of the couplers being too many for an otherwise exact-constraint configuration that resembles a four-bar linkage, which would only be applicable to limited-angle operation.

Rotary-to-Orbiting Drive
This mechanism consists of an eccentric shaft that supports through a bearing a disc with a number of equally spaced bores which simultaneously contact fixed pins of a smaller diameter to give the same eccentricity as the shaft, Fig. 7.2b. As the shaft rotates, the disc will orbit, without rotation. This mechanism is the basis for so-called planocentric drives (see Sect. 3.4.1). The drive is shown with six hole/pin pairs. With the eccentric shaft supported in bearings in a casing, the drive will have two moving parts, hence conceal five plurality overconstraints; five hole/pin pairs being too many for an exact-constraint configuration.

7.3.2 Angular Overconstraints

An angular overconstraint is concealed in an assembly when it does not permit exploiting an additional relative angular motion at some point in the assembly; orientation adjustment, as an expected consequence of inferior dimensional accuracy and/or deflection under load. One angular overconstraint renders the components incapable of stress-free loop closure in case of closed interfaces, or incapable

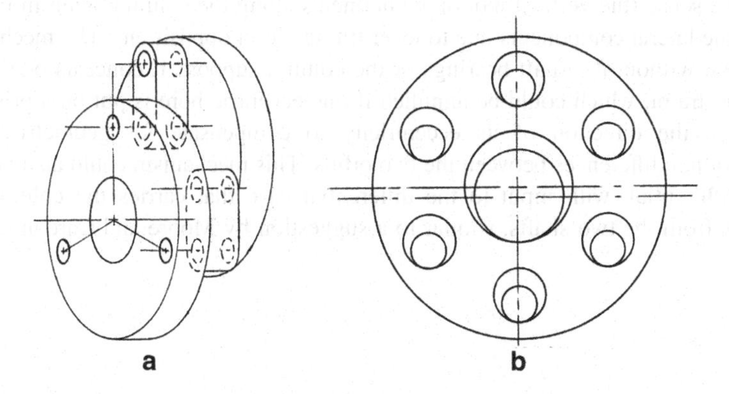

Fig. 7.2 Close-center drives, **a** multi-coupler drive, **b** planocentric drive

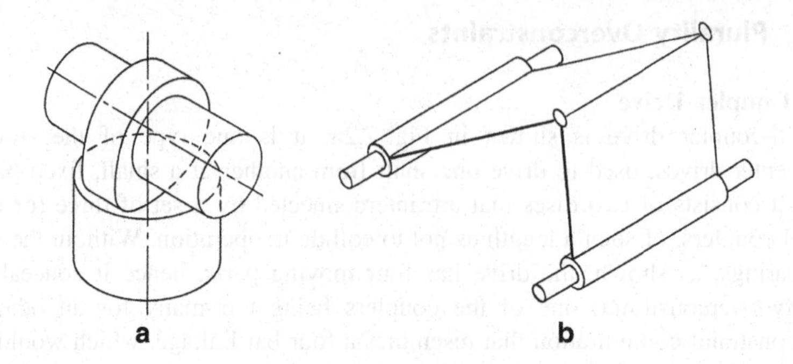

Fig. 7.3 Examples of mechanisms with alignment overconstraints, **a** cam on flat follower, with one overconstraint (shown without the camshaft bearings and the bucket tappet bore), **b** two R-hinged wings on two C-guides, with two overconstraints

of perfect seating on a straight-line or a flat surface in case of open interfaces. The latter condition results in edge loading. Figure 7.3 presents two examples of mechanisms concealing one and two angular overconstraints.

7.3.3 Lateral Overconstraints

A lateral or linear overconstraint is concealed in an assembly when it does not permit exploiting an additional relative lateral motion, axial or diametrical, at some point in the assembly, as an expected consequence of imperfect accuracy and/or deflection under load. A lateral overconstraint renders the components incapable of stress-free loop closure and the assembly incapable of smooth operation.

An example is given in Fig. 7.4 of a rotary motion inverter, which is based on a winged or cruciform unity-ratio lever that slides over a round column and carries two spherical-segment rollers, by which orbiting of an eccentric bore in one shaft end is mirror-imaged to orbiting of the eccentric bore in the other collinear shaft, in opposite sense (the vertical motion components along the column being in unison, while the lateral components due to lever tilting are in opposition). The mechanism is shown without the shaft bearings or the column support. It conceals one lateral overconstraint, which could be annulled if one eccentric bore is put on a prismatic guide in the direction of its eccentricity to compensate for geometrical and dimensional differences between the two orbits. This mechanism could also be used as a differential, with input to the differential case that carries the column and outputs from the two shafts, similar to a suggestion by Moore and Carden (1998).

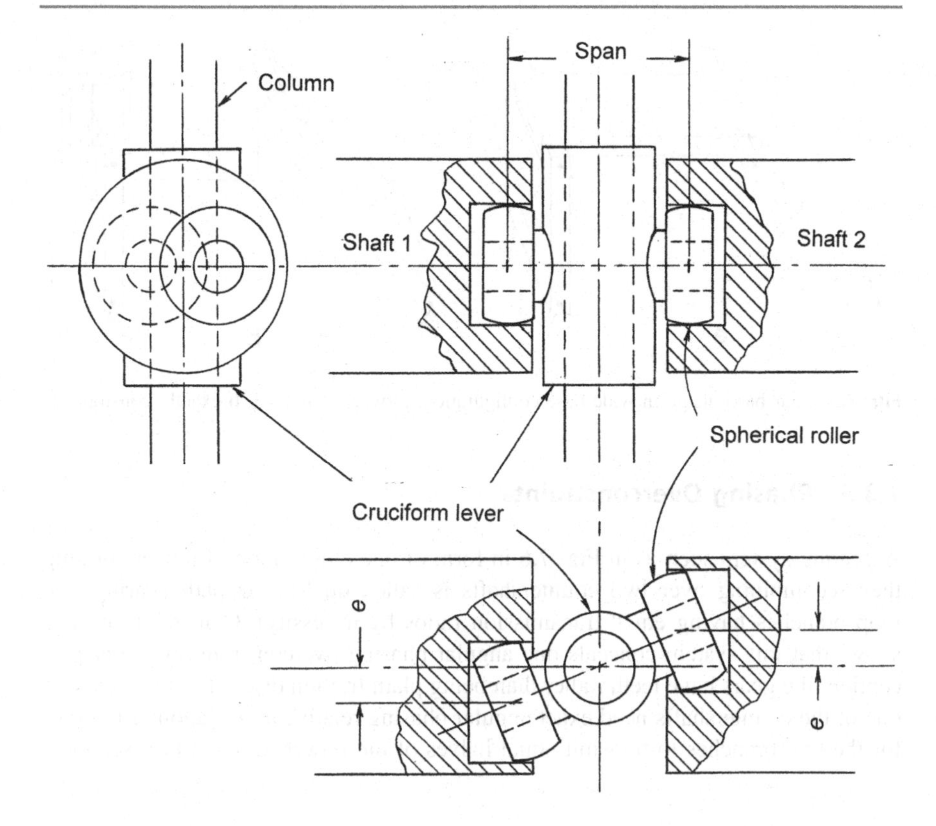

Fig. 7.4 Rotary motion inverter

7.3.4 Angular and Lateral Overconstraint Combinations

Four-bar linkages consist of an input crank or rocker, a coupler and an output rocker, all connected by revolute joints; 4R mechanisms. They represent a classical example of single-DOF mechanisms. They are planar in function but three-dimensional in reality, especially when they have to support out-of-plane loads. When all the links are of a wide-base configuration, such as in Fig. 7.5a, the assembly will be overconstrained: Grübler's count = $3 \times 6 - 4 \times 5 = -2$, which result represents the anticipated single mobility minus three overconstraints. The consequences of the latter are encountered as the last joint in the open chain is to be established; it will require the two links to freely align in two perpendicular angular directions, and to have lateral or end float along the pin axis. The mechanism thus conceals two angular overconstraints and one lateral overconstraint.

For a four-bar linkage to be exactly constrained, one solution is to give the coupler one more DOF in its connection at one end and two more DOF at the other. This is shown implemented in Fig. 7.5b as the coupler has a universal joint at one end and a spherical joint at the other.

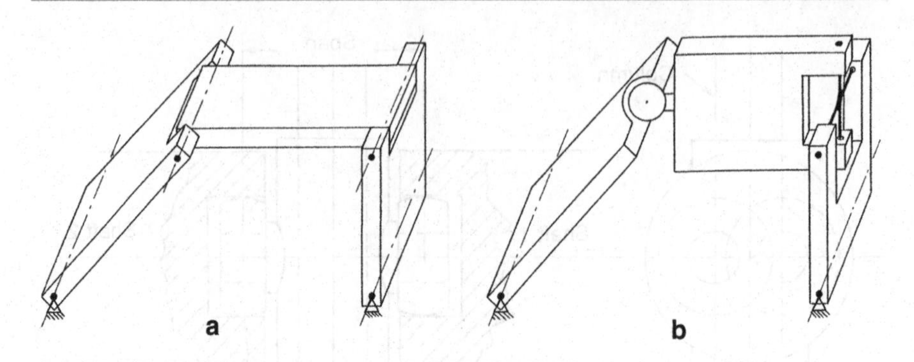

Fig. 7.5 Four-bar linkage in wide-base configuration, **a** overconstrained, **b** exactly constrained

7.3.5 Phasing Overconstraints

A gearing system such as in Fig. 7.6 in form of one closed loop of power splitting then recombining over two countershafts is called dual-power-path gearing (the loop branches having equal transmission ratios by necessity). Constraint analysis shows that this system conceals one angular phasing overconstraint in planar perception; the gears have teeth rather than being plain friction discs. The two gears on one of the countershafts need exact angular phasing relative to one another in order for the two branches to transmit equal halves of the power, or even to assemble.

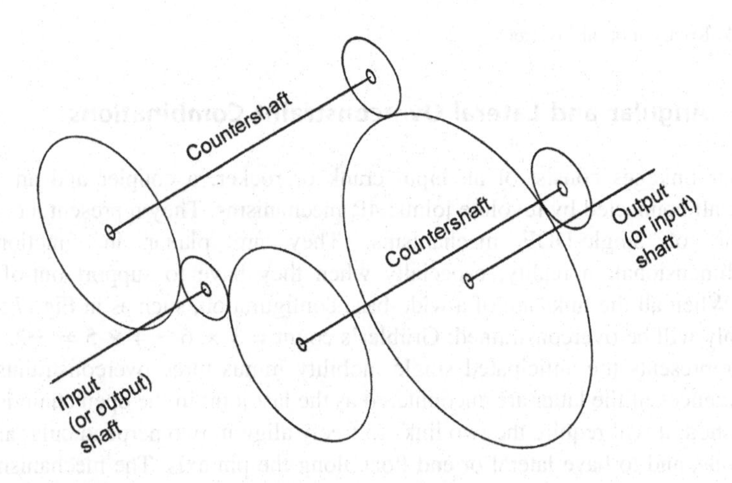

Fig. 7.6 Schematic of gearing of one closed loop; dual-power path gearing

7.3.6 Intrinsic Overconstraints

By default, a single-surface or a single-line contact between two rigid bodies is assumed to be of theoretically perfect geometry, giving an exactly constrained interface or joint; the assembly still being an open loop. The shortest closed loop is obtained with two rigid components in double contacts; each body having two separate features. The joint or the combined interface thus obtained will not—in itself—be exactly constrained anymore; it will be overconstrained. Such localized overconstraints may be termed *intrinsic*. Examples of two rigid components in double or multiple contacts, concealing intrinsic overconstraints, include the following.

1. Meshing gear teeth where two pairs of teeth are simultaneously in contact while traversing part of the path of contact.
2. A shaft with a journal and an adjacent shoulder in contact with a bearing sleeve to form a revolute joint.
3. A stepped shaft in a stepped sleeve.
4. Universal joint spider in the 2 × 2 yoke bearings.
5. Spline connections.
6. Two opposite, minimum-clearance contacts such as a flat tongue in a yoke or a helical pair when considering contact on both thread flanks.
7. A tapered part merging into a spherical head in contact with a cylinder while the taper wobbles in line contact inside the cylinder, such as in bent-axis pumps and motors of the tapered-piston type (see Sect. 11.2.2).

However, *obvious* geometric exactness of such geometric relations as collinearity, parallelism, perpendicularity, point and line coincidence, and the like could easily be assumed for the listed interfaces. This is particularly so since they are outcome of high volume production processes. Thus, the intrinsic overconstraints could be considered of only theoretical interest; ignored in the constraint analysis of larger-scale mechanical assemblies in order to sort out and concentrate on the overconstraints that really matter. This also applies to such machine parts as rolling-element bearings, Rzeppa joints, and the like.

7.3.7 Drawbacks Suffered by Overconstrained Designs

It takes a good deal of expertise to evaluate the consequences or adverse effects of overconstraints in a given design in view of their count and variety of types, but also in view of the accuracy of the inter-joint bodies (if any), of which the following definition is due. *Accuracy* refers to the closeness of the actual linear and angular dimensions of a machine part to their values specified in or even anticipated from the design drawing of that part. *Inaccuracy* is the countersense of accuracy. The adverse effects of the overconstraints could range between notably affecting the assembly and durability, and practically nothing at all. In general, the following drawbacks could be encountered.

In One-off and Small-batch Production

- When inter-joint bodies are manufactured with inferior accuracy, then the assembly of the rigid parts may not be completed at all. This will be perceived upon attempting to close at least any one of the loops/chains, being just a minor-closed loop between two bodies or a more or less extensive chain. With some compliance of the parts the assembly could be accomplished, but only a limited range of squealing operation will be achieved.
- When inter-joint bodies are manufactured with medium/acceptable accuracy, then the parts may assemble in only one particular position/orientation within the range of operation, but the mechanism will tend to squeal and jam when urged to depart from that point. These effects become aggravated under load due to the inevitable elastic deformation.

In Loaded Operation of Absolutely Accurate Parts

- Should the inter-joint bodies be manufactured with absolute accuracy, then the mobile assembly will function as anticipated, under no load, but the joints will suffer under load. Examples are some of the mechanical contrivances that prove to be of an unacceptably short service life, without an obvious reason.
- In case that unilateral straight-line contacts exist then edge loading would result, the severity of which being commensurate with the extent of inaccuracy and the heaviness of loading. Examples are spur and helical gearing with slender pinions, and straight-tooth bevel gearing without crowning.
- In case of multiple unilateral contacts in parallel, aiming at equal load sharing, the load will be transmitted by the number of contacts minus the number of overconstraints if the accuracy of manufacture was inadequate. One example is that of the required accuracy of angle phasing the intermediate shaft gears and pinions in split-power-path gearing; a rarely encountered necessity in the practice of gear manufacturing yet of grave consequences when neglected (see also Sect. 7.7).

Where Inter-joint Bodies do not exist

- The classical example is rolling-element bearings (especially of the single-row type), where the balls or rollers do not represent inter-interface bodies as such (between the races), and where they are finished to exactly the same diameter.

7.4 Identifying the Anticipated Mobility

Many machines and devices are of one-DOF by their very nature, engines and electric motors just for example. Kinematically flexible couplings between machine modules are available in a vast variety of types and sizes, and they are designed with one to five DOF to suit the various application circumstances. Identifying the anticipated mobility of these devices should pose no difficulty.

The anticipated mobility of certain other machinery may sometimes have to be carefully identified, not only for assessing the design being exactly constrained or overconstrained. One important category of machines that require such consideration is parallel kinematic machines (PKM), mainly used as manipulators or positioners. A simple example of such devices is the well-known 3-DOF manipulator of Fig. 7.7, which is capable of tilt in two mutually perpendicular directions (hence omnidirectional reach), as well as radial extension. It consists of a movable platform connected to a fixed base through three identical legs at 120° apart that each imposes one constraint on the assembly by being made of two members in a series RSR configuration, or an equivalent such as four members in a series 5R chain. In addition to being identifiable as a 3-DOF mechanism it is seen, by intuition, that the manipulator could be assembled in whatever position; it is an exact-constraint design. Three actuators will be required to independently position the three lower half-legs relative to the base.

Should an inventor be tempted to design a similar mechanism with four of those legs at 90° apart, for the apparent symmetry in executing the two tilting motion components, the PKM will conceal one overconstraint. Less-than-absolute

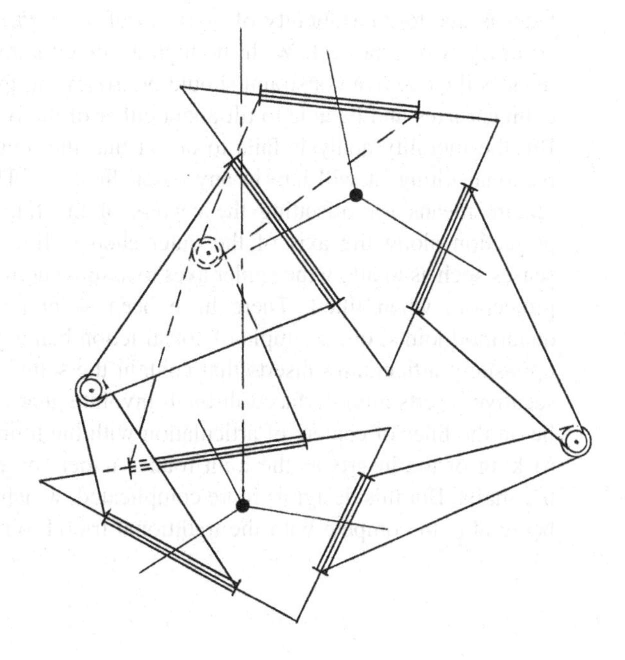

Fig. 7.7 Three-DOF manipulator schematic

manufacturing accuracy will make the mechanism jam, even under no load. But more importantly, three actuators are still necessary and sufficient to operate the manipulator; the suggested fourth one will make all four actuators *fight* against one another.

7.5 Limitations of Mobility Analysis

In a few situations the scheme of mobility analysis fails to detect certain particulars of the mechanical assembly, leading to wrong conclusions as to its being exactly constrained or overconstrained. These cases are presented in the following.

1. The fact that an exactly constrained mechanism could pass through a momentarily overconstrained situation as it moves. This case is encountered with parallelogram mechanisms; with two equal cranks and a coupler equal to the base which, if they were to simultaneously traverse the two dead centers, where all the links coincide. This category of undetectable overconstraints may be termed *simultaneous dead-center overconstraints*.
2. In some cases, two mutually perpendicular angular DOF do not represent two orthogonal components of omnidirectional mobility. This ambiguity is found when attempting to analyze a so-called quadripod joint, if constructed just as an extension of the well-known tripod joint, with a cruciform spider. In addition to plunging, such a joint allows the spider to tilt about either of its two mutually perpendicular axes. Constraint analysis gives Grübler's count of 6, viz., 3 anticipated DOF, 4 idle mobilities of the rollers, hence one overconstraint. The latter is due to the difficulty of insertion of a *fourth* roller if the manufacturing accuracy was imperfect. With the high accuracy known in manufacturing such devices the one overconstraint should be irrelevant; the joint will assemble with collinear axes and be able to tilt about either of the two spider axes; three DOF. But the mobility analysis fails to detect that this joint is incapable of omnidirectional tilting; it will jam in any other direction. The quadripod joint should feature means for offsetting the centers of the four spherical rollers—in the projection along the axis of the outer shell (tulip)—in successively opposite senses such as to allow the spider axes to assume non-orthogonal X-shape in that projection, when tilted. There have been several suggestions for designing quadripod joints; one example of construction being given in Fig. 7.8 with four oppositely articulating inserts that contain the semi-cylindrical races. The consecutive inserts are interfaced through involute gear tooth-in-space pairs, which lie on the lines of centers of articulation with the tulip to allow small-amplitude rocking of the inserts as the joint rotates under some operating angle between the shafts. But this design is more complicated, which may be the reason for not being able to compete with the traditional tripod joint.

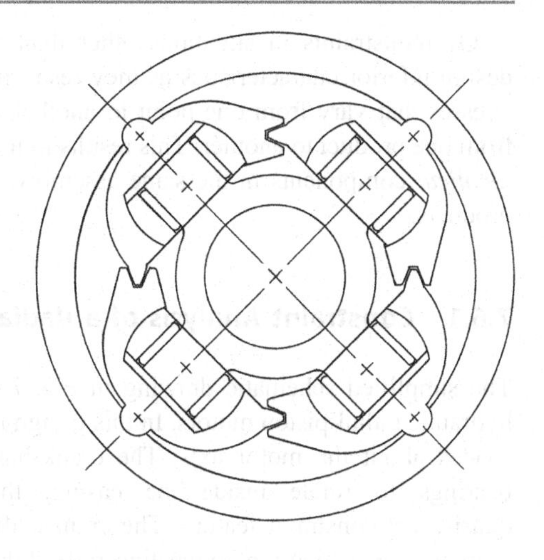

Fig. 7.8 Quadripod joint, shown in the position of collinear shafts

7.6 Quasi-Exact-Constraint Design

The kinematic constraint analysis of mechanical assemblies is an effective tool for deciding upon the design being exactly constrained or else overconstrained. This is an essential step toward assessing the functionality and, more importantly, the durability of the assembly, particularly if it operates under heavy loads and/or in a range of temperatures or temperature gradients. But not any design concealing overconstraints of any number or nature will be considered defective; many successful mechanical assemblies are not totally free of overconstraints, and they are considered the best possible solution to some engineering problems, just as they are. These can be called quasi-exact-constraint designs.

Distinct Cases of Quasi-exact-constraint Designs

1. Designs that conceal only one or two overconstraints of such a nature as to allow high manufacturing accuracy and finite rigidity of the assembly to effectively mitigate their effects, and where getting rid of (some of) them by kinematic means would be costly and/or cumbersome. Examples are heavily loaded linkage mechanisms, among several other devices.
2. Designs which are pending the validity of some simple assumption(s) regarding interface contact/mobility characteristics for them to be considered exactly constrained. An example is found in some types of curved-tooth gearing (see Sect. 14.4).
3. Designs featuring a plurality of intrinsic overconstraints owing to the existence of several identical, equally spaced items, which designs will not suffer any negative consequences if manufactured with utmost accuracy, especially where maintaining that accuracy is easy and straightforward as in mass production. This type of overconstraints even contributes to increasing the rigidity and load carrying capacity. Famous examples are the rolling-element bearings.

Overconstraints in situations other than those listed above would render the design inferior characteristics as they result in unpredictable amounts of locked-in stresses that vary from one point to another within the same assembly, as well as from one product to another. This results in inconsistent durability of the *apparently identical* components in the same assembly, and between the apparently identical products.

7.6.1 Constraint Analysis of a Radial-Piston Motor

The simplified schematic drawing in Fig. 7.9 depicts one typical construction of hydraulic radial-piston motors. In this design there are five pistons/cylinders equally spaced about the motor axis. The crankshaft is supported in two tapered roller bearings to rotate inside the casing, thus emulating a revolute joint; a quasi-exact-constraint feature. The crankshaft has an integral eccentric cylindrical journal, over which the connecting rods slide with their partial-cylindrical bottom surfaces, without being constrained in the lateral (axial) direction. The connecting rod upper ends are spherical in shape, and they fit inside spherical seats in the pistons. The drawing does not show the hydraulic distributor at the rear end of the motor.

Constraint Analysis

1. Number of moving parts: 1 crankshaft, five connecting rods, five pistons = 11. Total degrees of freedom of the parts before being put in the assembly = 6 × 11 = 66. Interface constraints imposed owing to assembly: 1R, 10C, 5S giving 5 + 40 + 15 = 60

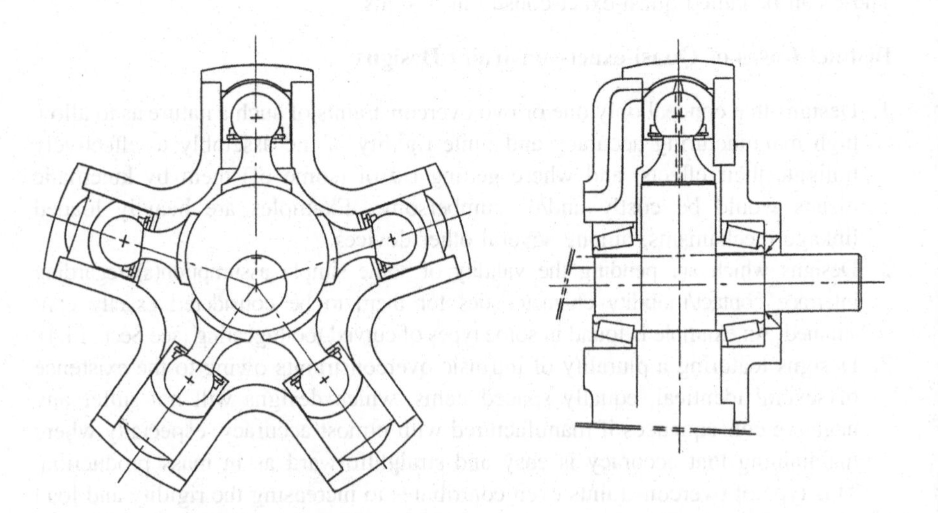

Fig. 7.9 Hydraulic radial-piston motor

Grübler's count = 66 − 60 = 6.

Anticipated mobility of the motor = 1.

Idle/independent mobility = 5; the possibility of the pistons freely, slowly rotating about their own axes.

AMIGO equation: 1 + 5 = 6 + 0.

The assembly is exactly constrained.
2. The motor should perform smoothly, without any possibility of jamming.
3. Therefore, this motor type belongs to the so-called surviving, well-proven designs.

7.7 Pitfalls in Designing Locked-Train Gearing

Gearing systems that consists of at least one closed loop of power splitting then recombining (over countershafts) are called locked-train gearing. This is because they conceal phasing overconstraints (see Sect. 7.3.5). Ignoring these overconstraints would represent a serious design pitfall. But their detrimental consequences could be evaded or mitigated with any of the following measures taken to ensure equal load sharing by the branches of the individual loop(s); for the system to deserve other names such as power-split gearing (see Sect. 9.2).

1. Phasing/timing adjustment of the two gears relative to one another in one branch (on one countershaft) with sufficient, equal torsional compliance being provided in both branches, e.g., through quill–torsion shaft combinations to deal with any residual inaccuracies of phasing (see also Sect. 8.3.1).
2. Making the high-speed stage consisting of two single-helical gears of the same helix angle but of opposite hand, which are disposed in two parallel offset planes to mesh with one axially floating twin-helical pinion. This pinion will adjust itself to be in equilibrium under two equal and opposite axial load components, hence to transmit two equal tangential loads. Result is equal power split to the two gear meshing contacts, which will be maintained on the two sides with no possibility of there being overconstraints. The countershafts should be axially restrained, either by thrust bearings or by making the low-speed gear set double helical. This measure is according to a basic old principle introduced by Alquist (1920).

The same measure as of (2) is also suitable for gearing systems consisting of a number of multiple identical loops in parallel, where the single loops are again locked with one another by driving the same number of double-mesh pinions, in a circular array. Such a system will be named *interlocked gearing*. An example is the

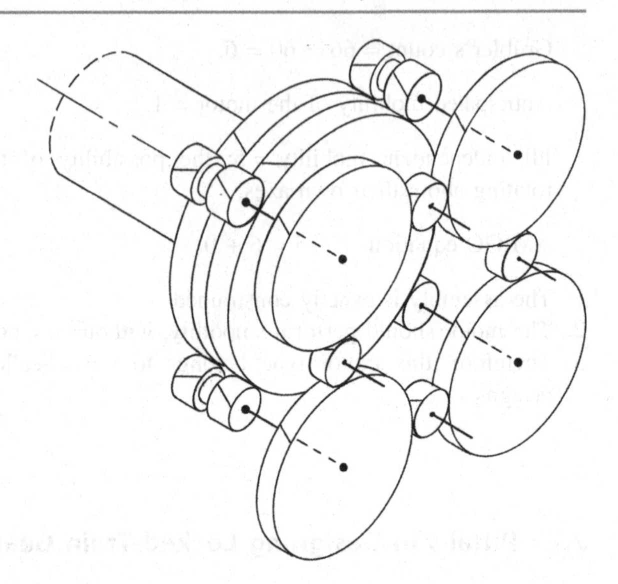

Fig. 7.10 Interlocked step-up gearing with four outputs

gear train shown in Fig. 7.10, which is similar to that by Mikhail et al. (2006); intended as a wind turbine transmission. It consists of a double-helical input stage and a coplanar, single-helical, high-speed stage with four output pinions to drive four individual generators or hydraulic pumps. However, it does not apply any of the above-mentioned measures for load sharing equality. The resulting overconstrainedness and its consequences are discussed in the following.

Constraint Analysis This gear set could simply be treated as a planar mechanism (the second-stage gearing being helical alters nothing regarding assembly), otherwise a large number of overconstraints would result just due to the fact that neither the parallelism of the nine shafts nor the equality of the helix angle on meshing gears is taken for granted. These two issues being easily maintained with high accuracy, the sought overconstraints; the root cause of failure should lie elsewhere. Therefore 9 moving parts of 27 DOF before assembly, 9 revolute joints imposing 18 constraints, and 4 tooth flank meshes in the first-stage and (presumably) 8 in the second-stage imposing 12 constraints. The total number of constraints due to assembly is 30 and Grübler's count becomes −3, instead of 1. The assembly is thus fourfold overconstrained; it conceals four phasing overconstraints.

The Overconstraints Revealed and Interpreted These overconstraints surface in the last assembly steps as each of the four high-speed pinions is to be brought into mesh with its two gears simultaneously, which would probably not succeed. After tedious trial and error of relocating the countershafts the pinions may, at best, go into single-drive-flank mesh with one of the two gears, while backlash in the opposite tooth/space will be divided into coast-side backlash and drive side flank separation or *mismatch* ranging anywhere between zero; the rare ideal case for equal load sharing under balanced tangential loads, and the full amount of backlash. In any but the ideal case, with any amount of backlash mismatch the load will be carried on one drive side. Having lost four mesh points the number of constraints will be reduced by four, which results in a Grübler's count of unity, and the assembly would again become exactly constrained, but only fictitiously; if the gearing was intended for single-drive-flank operation, which is not the case.

Consequences The consequences of single-drive-flank contact, as opposed to proper double-meshing conditions as anticipated, are (1) doubling the tooth tangential load; much reducing their durability, and (2) acting as a radial load on the output pinion (instead of being in equilibrium under two equal and opposite tooth loads), bending the pinion shaft, which results in severe tooth edge loading. Tooth fracture will be imminent on one of the pinions, first. It is therefore seen that overconstraints could be very injurious to the gearing system, particularly with ample backlash being specified and with rather short (cantilever) output shafts. Actual premature failure cases of wind turbine gearboxes of the present configuration were reported by De Vries (2008), for example.

In case the backlash was held to the minimum recommended amount and the output shafts were made long enough, the system would perform *better*. However, according to the amount of backlash mismatch (between zero and maximum) the extent of output shaft bending and, hence, tooth edge loading would differ between the individual pinions. The gear/pinion tooth flanks would thus be of inconsistent durability.

Implication of the overconstraints is that the system, in its present configuration and to operate and be loaded as anticipated, should have been of absolute exactness regarding computing, specifying and actually achieving angular phasing or timing of the pinion/gear on the individual countershafts. However, timing of gear teeth on one shaft is something the manufacturers are not accustomed with, and it would require an inventive timing metrology fixture to assess the exactness of the procedure prior to accepting the product.

7.7.1 Interlocked Gear Train with Equal Load Sharing

Split-power-path gearing could be made quasi-exactly constrained when configured as in Fig. 7.11. In a two-stage step-up gearing with input to a spur or slightly helical bull gear that drives, say, six equally spaced first-stage pinions which are mounted on axially restrained countershafts, which also carry the second-stage helical gears, the latter are made in an axially staggered layout and of alternately opposite-hand helix, each two consecutive ones to drive an axially floating, double-helical (twin-helical) output pinion to equalize the loads according to the principle originally introduced by Alquist (1920), every three shafts being coplanar (see Sect. 7.7). The six outputs may be intended to drive six individual electric generators or hydraulic pumps. For assembling this interlocked gearing configuration, each output pinion should be tilted in the common plane of the three axes, inserted to its approximate location, leveled to mesh with its two helical gears with some angular/axial manipulation, before its two bearings could be inserted and fastened.

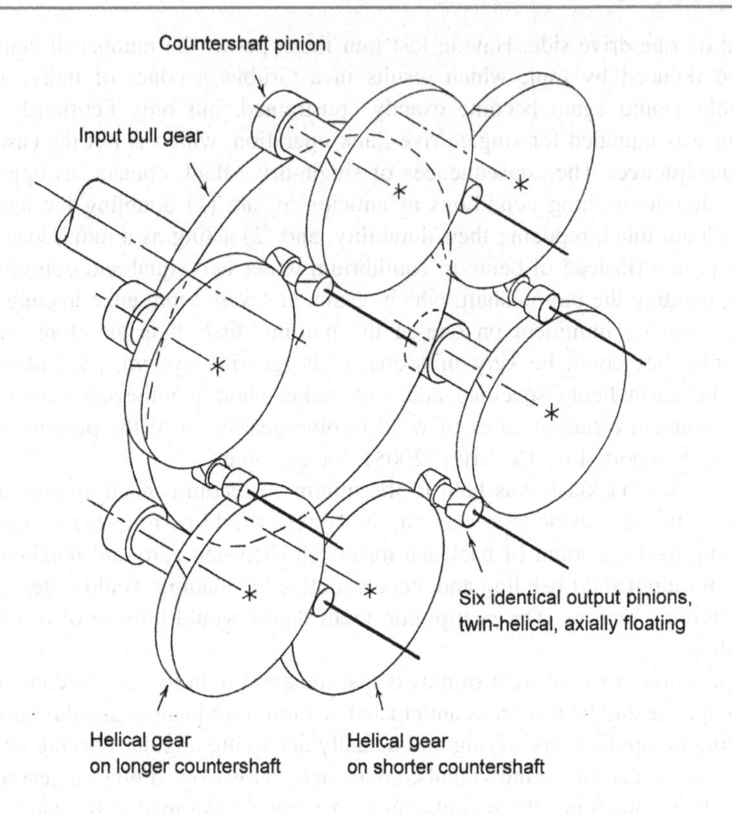

Fig. 7.11 Six-output interlocked gear train with equal load sharing

7.8 Spur Gears Made Tolerant of Misalignment

Unlocking the one relative motion constraint that prevents skewing in or parallel to the plane of action could make spur gearing a good example of self-aligning designs. There exist two methods for providing full-face bearing/contact.

Method 1 Supporting the pinion bearings in a yoke or cradle that can pivot about an axis in the middle transverse plane of the gears, perpendicular to the plane of action, as shown in Fig. 7.12. The normal tooth load, transmitted as torque plus the same load at the pinion center, should be directed from that center to the mid-span of the cradle articulation pin. This renders the pinion insensitive to misalignment in the plane of action, while retaining its capability of slightly adjusting in the common-tangent plane. The pin should be disposed as close as possible to the pinion, and should be of sufficient length to deal with any inadvertent load changing situation. The limitations of this construction are that it is only usable in unidirectional loading (as indicated on the gear if driving), and necessitates that the power to/from the pinion be transmitted through a 5-DOF drive

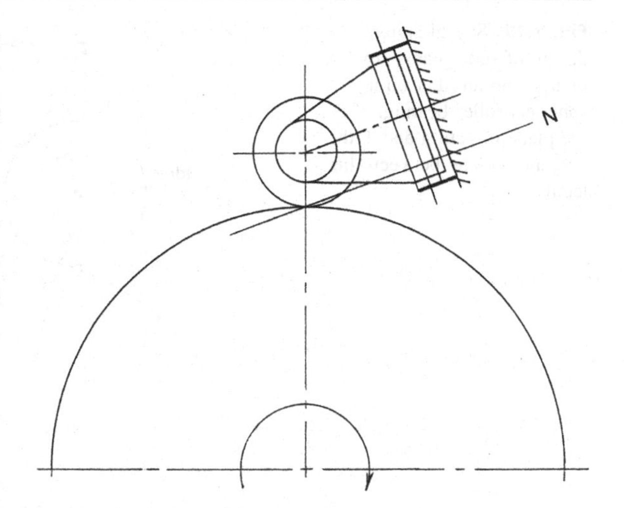

Fig. 7.12 Self-aligning design of spur gearing by supporting the pinion in a cradle that is pivoted to align the tooth contact lines in the plane of action N

shaft such as a plunging double Cardan shaft, since the pinion should be free to move in whatever direction as to align to full-face bearing/contact.

Method 2 Mounting the gear and pinion on fixed axes, pulled apart to interpose an idler in such a location as to make two mutually perpendicular planes of action (N) with the two gears (the two tangent planes T will also be mutually perpendicular, and the angle included between the two lines of centers will be 90° + twice the pressure angle φ), Fig. 7.13. The idler should be supported on a spherical roller bearing, or more appropriately on a (slightly) toroidal roller bearing with a large width-to-diameter ratio. The idler tooth flanks can thus adjust to full-face bearing/contact on the one flank interface without affecting the contact on the other, and vice versa. The advantage of this construction is that there are no special requirements on the shaft coupling to transmit power to/from the pinion. Drawbacks are that the design is only usable in unidirectional loading (as indicated on the gear if driving), includes a larger number of components, needs a large idler to accommodate the necessary bearing size (load = $\sqrt{2}$ the normal tooth load F_{n}), and results in larger radial dimensions of the gearbox. Loading in the opposite sense does away with the attribute of self-alignment about two mutually perpendicular axes, in addition to placing the idler at its wrong location; to be under two almost parallel reactions to add up to a bearing load of about twice the normal tooth load. The basic idea is disclosed in the patent by Quenneville (1969).

In summary, neither of the two designs discussed in this Section seems to represent a most satisfactory solution to the problem, such as the one offered by curved-tooth gearing (see Sect. 14.4).

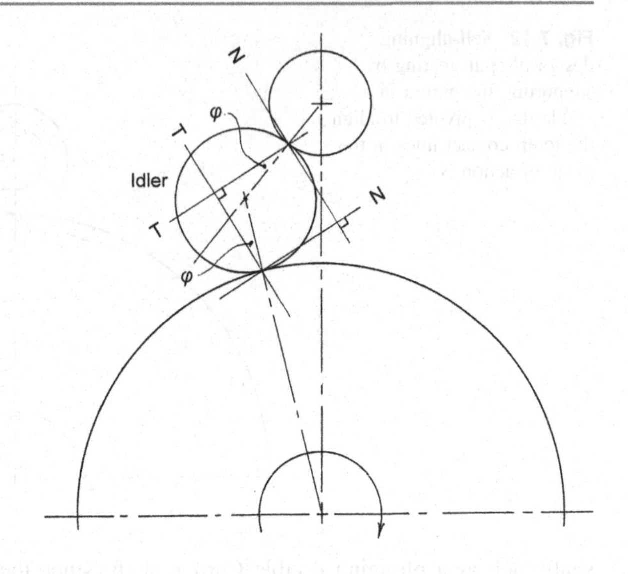

Fig. 7.13 Self-aligning design of spur gearing by interposing an idler on a spherical roller bearing. N is the plane of action and T the tangential plane between the teeth

References

Adrian W (1926) Tagung für Getriebelehre in Dresden im Oktober 1926. ZAMM 6(6):487–494

Alquist K (1920) Gearing. U.S. Patent 1,351,317

De Vries E (2008) Launching liberty: clipper addresses teething issues. Renewable Energy World, 22 July 2008

Gogu G (2005) Mobility of mechanisms: a critical review. Mech Mach Theory 40(9):1068–1097. https://doi.org/10.1016/j.mechmachtheory.2004.12.014

Grübler M (1909) The criterion for constrained motion of a system of screws. Paper 9 in: Dintzi E (1909) The Salzburg meeting of the Deutsche Mathematiker-Vereinigung. Bull Amer Math Soc 16(3):114–121

Mikhail AS, Hahlbeck EC (2006) Distributed power train (DGD) with multiple power paths. U.S. Patent 7,069,802, 4 July 2006

Moore JW, Carden JC (1998) Differential drive mechanisms. EP Patent 0 611 166 B1, 14 January 1998

Quenneville RN (1969) Gear alignment means. U.S. Patent 3,434,365, 25 March 1969

8 Allocation of Functions

The basics of active redundancy, implemented whenever deemed appropriate for raising the reliability of a mechanical system, are explained with various examples on the necessity of provision for autonomous isolation or decoupling of any of the redundancy components upon failure. A real-life case study of a hydraulic system subjected to the hazard of high-ratio pressure intensification is presented to show how ignoring the importance of redundancy in particular situations could lead to system destruction. The principle of the separate allocation of functions is explained and its necessity in some cases for improving the system durability is exemplified through some types of multi-stage gearing. Examples on the proper and improper association of functions are given; the latter involving the risk of running into the pitfall of conditional functionality.

8.1 Allocation of Functions to Their Carriers

Any machine component is there to fulfill a functional requirement; to realize a *function* for short. Accordingly, that component is generally referred to as a *function carrier*. The allocation of functions to function carriers could fall under any of the following categories.

1. A single function, being allocated to one function carrier; a simple case that needs no further explanatory examples.
2. A single function, being allocated to two or more function carriers, identical or non-identical, that act in parallel. This is the definition of *active redundancy*.
3. A set of distinct functions, usually dual or triple functions, pertaining to one place in the system, being allocated to separate function carriers. This is called separate allocation of functions.
4. A set of distinct functions that pertain to one location in the assembly, being allocated to a single function carrier. This is the case of association of functions.

H. A. Arafa, *Design for Durability and Performance Density*,
https://doi.org/10.1007/978-3-030-56816-0_8

Design decisions regarding any of these categories should be made with the principle of *proper allocation of functions* in mind. In particular, decisions regarding categories 3 and 4; the separate allocation of functions and the association of functions should be made with the utmost care (see Sects. 8.3 and 8.4).

8.2 Redundancy

Redundancy in mechanical engineering pertains to a function that could be carried by one unit or element, when it is assigned to two or more identical elements, of the same or smaller size, to function in parallel, under the imperative condition that any of them will autonomously be isolated or decoupled upon its failure, so that the remaining redundant elements could continue to function safely. Parallel redundancy and active load-sharing redundancy are more precise names, since all the system components are equally loaded in operation, also equally overloaded after failure of one. Dual and triply redundant systems are common-place, but a larger number of elements (N) could be involved in a redundant system. However, not any multi-element design should be indicative of redundancy; load sharing and a more compact design space would be the only benefits in such cases as multi-disc clutches, multi-strand roller chains, or planetary gear sets. Application of the principle of redundancy has one or both of the following objectives:

1. Using a number of smaller devices rather than a large one, in situations where it is known that performance density will be higher, and or for better maintainability.
2. Raising the system reliability (R), in a safety–critical situation, by duplicating or triplicating some components. Assuming no common-cause failures, such as power supply disrupt, the (small) failure probability of the system equals that of one element (F) squared or cubed, respectively. Since $R = 1 - F^N$ then the system reliability will be markedly increased.

8.2.1 Isolating or Decoupling Failed Components

In active-redundancy systems special attention should be paid to the condition that a failed component must be prevented from interfering with the operation of the surviving ones, or of the system as a whole. Therefore, such systems require an effective decoupling, isolating, or disengagement mechanism for the failed component; otherwise not much reliability enhancement could be expected from the application of the redundancy principle. In many cases complete decoupling between the redundant components is provided by the very nature of the system design. In some other cases special care has to be taken to ensure safe disengagement of a failed component. In a few cases dedicated decoupling mechanisms or elements have to be integrated into the system. Negligence of doing such is considered a design pitfall. It could thus be seen that redundancy sometimes produces less, instead of greater reliability. Redundancy comes at a cost; it increases system complexity, and should not be implemented unless justified.

8.2.2 Examples of Isolating Failed Components

1. **Twin-engine Helicopters**

The propulsion and power transmission system of twin-engine helicopters is a famous example of dual redundancy; the probability of two redundant components failing during a critical time period is much less likely than that of one component failing during the same period. The power of each engine should at least be sufficient to land safely, or even to continue the return mission.

Isolating Means The input from each engine to the gearbox should be through a highly reliable one-way clutch for safe disengagement of either engine when stopped or shut-off, as well as for independent engine starting.

2. **Nested Helical Springs**

Older designs of piston-engine valve mechanisms use two concentric helical springs for loading each valve; in an active load-sharing redundancy. The identical thing about the springs is equal stiffness and preload. The result is enhanced reliability; fail-safety should one spring breaks.

Isolating Means The springs should be of opposite hand for a broken one not to get entangled in the winding of the other.

3. **Hydrostatic Bearings**

Hydrostatic multi-recess bearings such as for the turntable of large vertical-axis machine tools feature a large number of recesses (typically divisible by 4) and are supplied with a number of identical motor–pump units in active load-sharing redundancy. The pumps are typically four-gear pumps, each having four independent constant supply-flow outputs to connect with four recesses that are disposed at $4 \times 90°$ as shown in Fig. 8.1 with four pumps. This arrangement provides even load distribution if one of the motor–pump units failed. This system is of quadruple redundancy. But the 16 fluid lines do not represent a redundancy, since failure of any line would necessitate switching off its motor–pump unit.

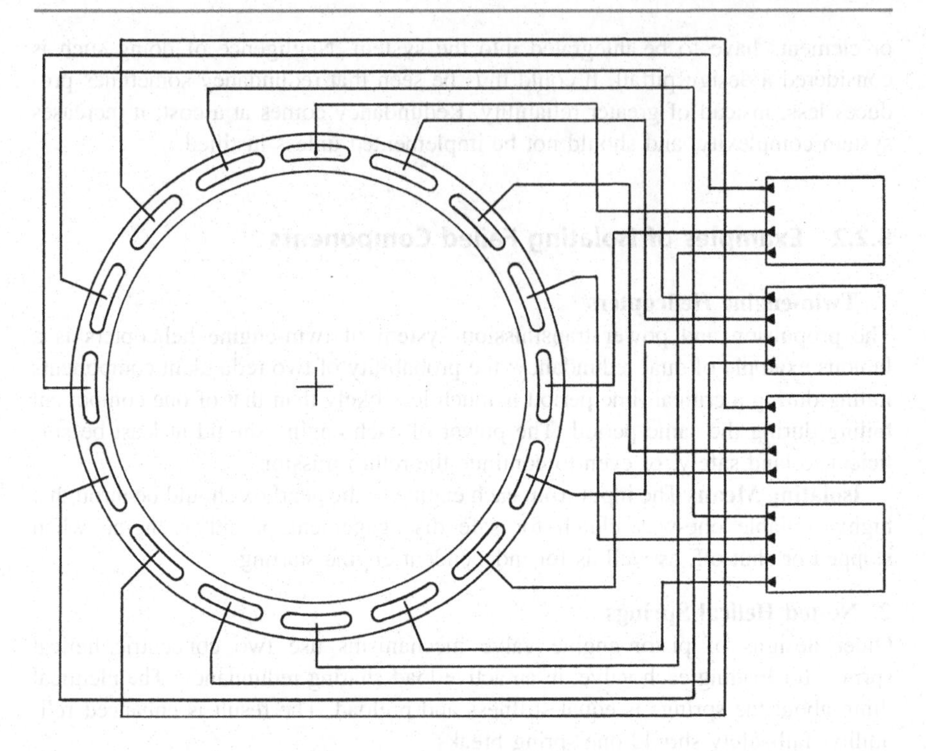

Fig. 8.1 Hydrostatic bearing with 16 recesses, each receiving a constant supply-flow rate, from four quadruple-output pumps

Isolating Means None; isolation is provided by the very nature of the design.

4. **Multiple V-belt Drives**

A widely applied method is to transmit larger powers with a number of the more flexible, smaller-section V-belts, in active load-sharing redundancy, instead of one thick belt. This gives a compact design and easier maintainability.

Isolating Means None, but abundant free space should be provided for a ruptured belt not to get entangled in the remaining ones.

8.2.3 Case Study

High-ratio hydraulic intensification

Hydraulic presses for sheet metal deep-drawing operations invariably use differential cylinders with a large area ratio. Following a fast *prefill* part of the downward extending stroke the cylinder should provide a large force by the system pressure acting on the full piston area in a slower working stroke, then a small uplifting force in a fast retracting stroke. The present case study deals with an unexpected, destructive failure of a hydraulic press with a differential cylinder of a large area ratio, an initial account of which case was given earlier (Arafa 2006). The construction drawing of the cylinder is shown in Fig. 8.2.

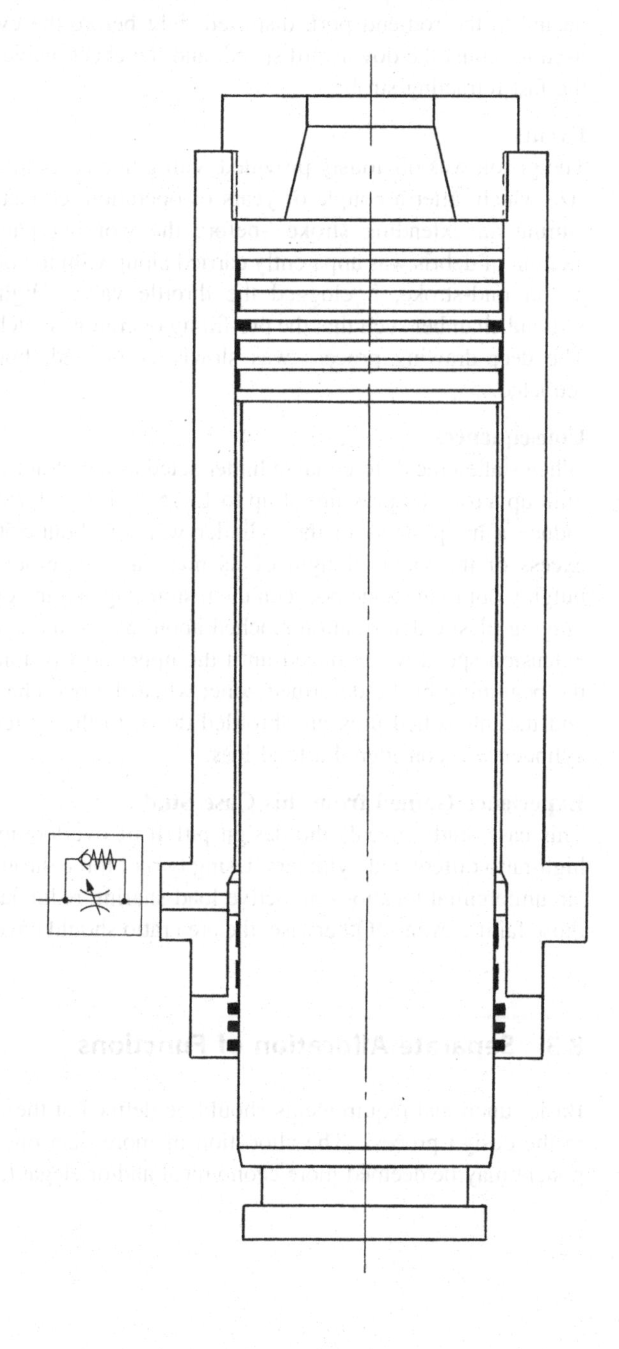

Fig. 8.2 Hydraulic differential cylinder with a large area ratio (scale schematic drawing)

Essential Data
Cylinder bore 540 mm, piston rod diameter 520 mm (giving an unusually high area ratio of 13.75), cylinder wall thickness 87.5 mm, system pressure 130 bar, hence the press capacity is about 300 ton.

The cylinder is provided with one adjustable throttle/check valve directly connected to the rod-end port, disposed right before the cylinder cap. The throttle is used to adjust the downward speed, and the check valve opens the fluid passage in the fast retracting stroke.

Events
The piston was obviously provided with guide rings that released debris of visible size which, after a couple of years of operation, collected in the cylinder bottom. During an extending stroke—before the working part of the process—a large amount of debris was apparently carried along with the out-flowing oil and, at about piston mid-stroke, it clogged the throttle valve. High-pressure built up in the rod-end chamber such that the pre-filling operation switched to full system pressure. The deep-drawing process was slowly completed, but the piston could not be retracted.

Consequences
When stalled the differential cylinder acted as a hydraulic intensifier that could have built up a rod-end pressure of up to $13.75 \times 130 = 1{,}788$ bar. This pressure would induce a hoop stress in the cylinder wall of about 550 MPa, which is much in excess of the yield strength of its material. Therefore, the cylinder wall started bulging out in the zone between the piston ring and the cylinder cap. The maximum bulging plastic deformation reached about 5% of the cylinder bore, and the piston extension speed was reduced until the uppermost piston guide ring crept down to the beginning of the deformed zone; when the two chambers were interconnected and the intensified pressure throttled down to the system pressure. The hydraulic cylinder was considered a total loss.

Experience Gained from this Case Study
This case study reveals the design pitfall of overlooking hazards associated with high-ratio differential cylinders. Using two or three throttle/check valves at different circumferential locations, in active load-sharing redundancy, could have prevented those failure events. Otherwise, the area ratio should have been made much smaller.

8.3 Separate Allocation of Functions

Basic functional requirements should be defined at the concept-development stage in the design process. The allocation of more than one function to a single component may be deemed more economical and/or elegant; keeping the design simple and compact by reducing the number of components. This decision may be referred to as functional association, combination, or integration.

However, there are situations where each component should be allocated a clear-cut functional requirement. This may be referred to as the dissociation, distinction, division, partition, or separation of functional requirements or *tasks*, which may be unified under the collective title of *separate allocation of functions*. Adopting this practice becomes particularly important in the following cases.

1. **Inadequacy of one component to fulfill two or more distinct functional requirements**. Example: Internal-combustion-engine pistons have to fulfill the two requirements of motion guidance inside the cylinder with active and reactive force transmission, and sealing the high-pressure gas. In reciprocating engines the first task is assigned to the piston itself, while the second task is taken by a set of piston rings. In rotary-piston engines the apex seals are not adequate for both functional requirements of sealing the three chambers off each other and guiding the triangular rotor in its motion inside the epitrochoid-lobed casing, taking the reactive loads. The latter functional requirement is assigned to an internal gear set of a ratio of 2/3.
2. **Conflicting or contradicting functional requirements; where simultaneous optimization for both is not possible**. Example: Power-shift drives of machine tools include a number of clutches for selecting the different gear ratios. The braking function of this drive train could be done by engaging two widely spaced speeds. Nevertheless, it is much preferred to have a separate brake for the deceleration and stoppage cycles. In this case the clutches could be appropriately sized for their function, and the brake could be placed outside the gearbox for better heat disposal and for better accessibility for inspection and replacement of friction pads or discs.
3. **Functional interaction**. Example: Hydrostatic bearings are often designed to carry a combination of radial and axial loads. A double-tapered bearing would seem to be simpler and less costly. Yet there will be a fixed ratio between the axial and radial load capacity. Separate allocation of functions will decouple this implication; a purely radial bearing and a purely axial one.

8.3.1 Examples of the Separate Allocation of Functions

Separate allocation of functions is typically applied in large gearboxes, where each gear has to have dual-functional requirements satisfied; those of

1. Supporting: precisely locating the gear in the transverse plane to maintain the pre-scribed center distance with its mating gear/pinion, and reacting the normal tooth load and the tilting moment in case of single-helical gears without thrust collars.

2. Torque transmitting: with an objective of minimizing misalignment-induced moments and their effect on the tooth-bearing pattern.

Pahl (1973), then Pahl and Beitz (1984) explained the "principle of the division of tasks for distinct functions" as applied to a large steam turbine gearbox that consists of a single-stage, single-helical gear pair, a design of the late 1960s. The bull gear is supported on a "stiff" quill shaft in a pair of journal bearings at "shortest possible span," while torque is transmitted from the far end by a radially and torsionally compliant torsion shaft that runs through the quill shaft. Figure 8.3 is a schematic of the bull gear support and connections in that design; a quill–torsion shaft combination. The gearbox is thus relieved of the effects of external misalignment, thermal expansion, and foundation settling. Inadequacy of the journal bearings with their substantial clearance, when mounted at a short span, for reacting the tilting moment on the gear by virtue of its being helical was, however, not commented. Neither was the fact that the bull gear is kept unnecessarily floating in axial direction, to be correctly located only upon installing the gearbox and coupling the output shaft with the next machine module.

Dudley (1962) briefly described the application of the same principle (without naming it) to each of the four countershafts of a large split-torque, two-stage, double-helical marine gearbox of a power rating of about 37 MW from two inputs. One of the countershafts is shown schematically in Fig. 8.4, on which the separate first-reduction gear and the second-reduction pinion are supported on their quill shafts in their own pairs of journal bearings at an ample span, and through which quill shafts a slender torsion shaft runs to connect their outboard flanges by being

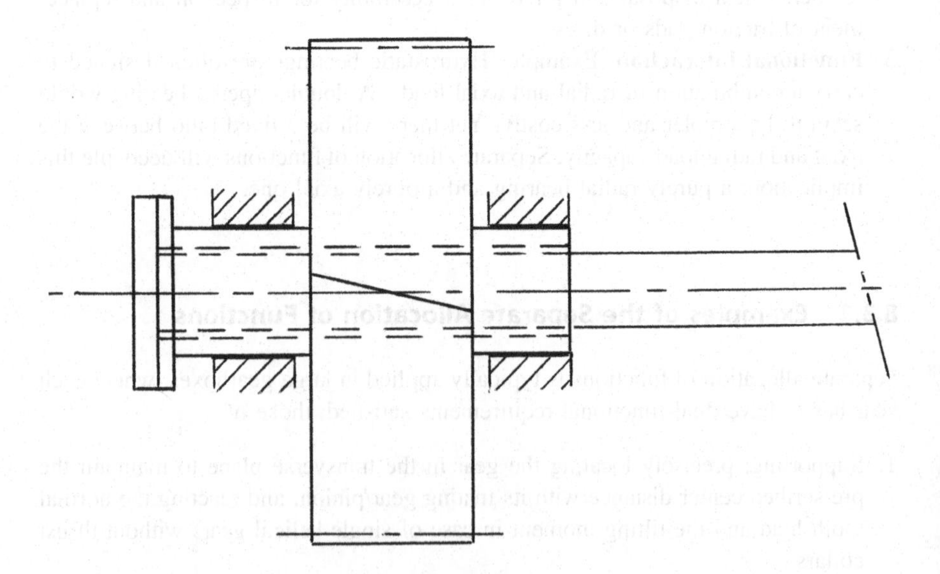

Fig. 8.3 Single-helical bull gear support and connections

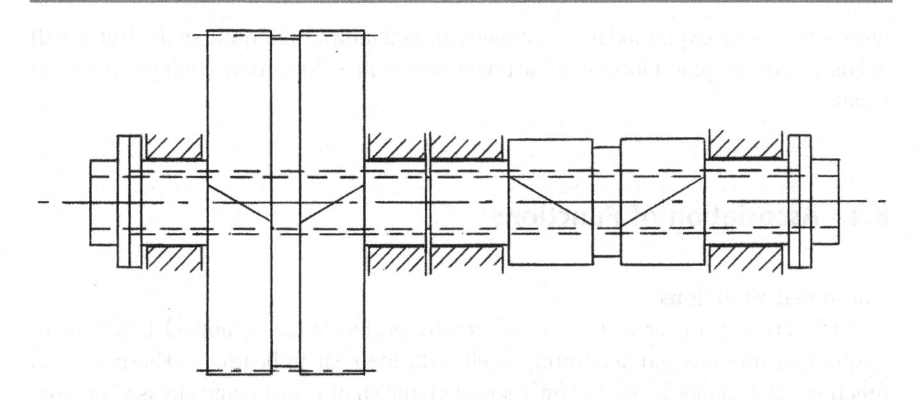

Fig. 8.4 Countershaft of a two-stage reducer with the double-helical gear and pinion separately supported on their quill shafts and connected by a torsion bar

splined and fitted in its far-end flanged hubs, in a fine drive fit. This arrangement is called an *articulated design*, a well-proven design in marine propulsion engineering. However, each torsion shaft has to be precisely adjusted in length between the two flanges to secure equal bearing on both halves of each double-helical component, and in phase to provide equal load sharing in the split-torque configuration. This gearing is thus highly overconstrained; it requires highly precise adjustments to conceal the effects of its overconstrainedness (see also Sect. 7.7).

A better alternative of using the quill–torsion-shaft combination is in the countershaft of two-stage gearing having a slightly helical high-speed stage and a double-helical low-speed stage, Fig. 8.5. The latter could also feature a little larger face width of one of the half toothings to react the single-helical axial thrust with equal unit loads in both halves. This construction is free of axial overconstraints; it

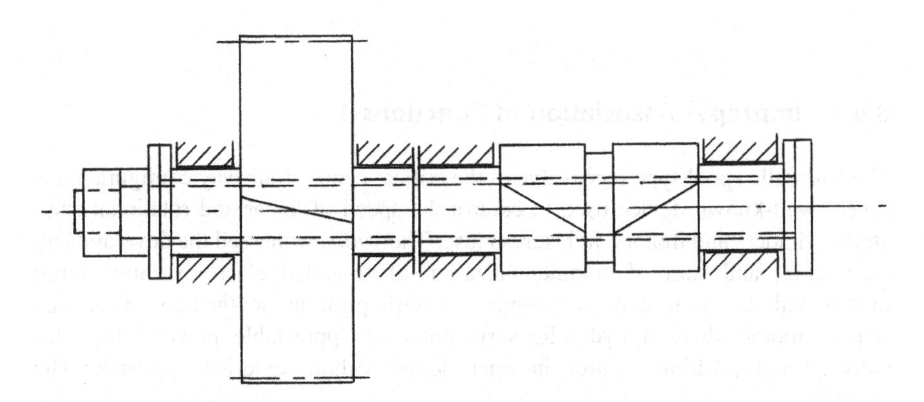

Fig. 8.5 Countershaft of a two-stage gearing with the slightly helical gear and the double-helical pinion separately supported on their quill shafts and connected by a torsion shaft

does not need accurate axial adjustments in assembling the quill shaft. But it still needs pinion-to-gear phasing adjustment when in split-torque configurations, as usual.

8.4 Association of Functions

Combined Functions
A shaft carrying a gear or the like inherently performs the combined functions of torque transmission and supporting itself with the gear in bearings. These are two functions that cannot be said to be associated; the shaft is just being stressed in more than one mode.

Associated Functions
The once highly appraised invention of associating the functions of *piston stroking* and of *synchronously driving the cylinder block from the drive shaft*—in bent-axis pumps and motors—within one function carrier (the group of tapered connecting rods or pistons) after Thoma et al. (1939) resulted in a highly overconstrained assembly that requires absolute accuracy of manufacture to mitigate the effects of the overconstraints (see Chap. 11). This design is now being gradually replaced by exact-constraint designs incorporating the principle of separate allocation of functions, where the pistons are relieved from the driving function, which is done by a pair of bevel gears in fixed-displacement units or a double Cardan or a double tripod shaft in variable-displacement units.

Further Examples
Design examples of flawless, advantageously associated functions are not many but instead, the *improper association of functions* will be highlighted by the two examples to follow.

8.4.1 Improper Association of Functions 1

The hydraulic spool-type servovalve of the critical-center, four-way configuration is a very well-known device used to control the speed of linear and rotational actuators; cylinders and motors, in fast response. The 1960s witnessed the development, maturation, and start of commercialization of so-called electrohydraulic pulse motors with an aim to make a low-power electric pilot motor (then conceived as a stepper motor) drive a hydraulic servomotor of appreciable power rating, for velocity and position control in open loop, without external feedback. The

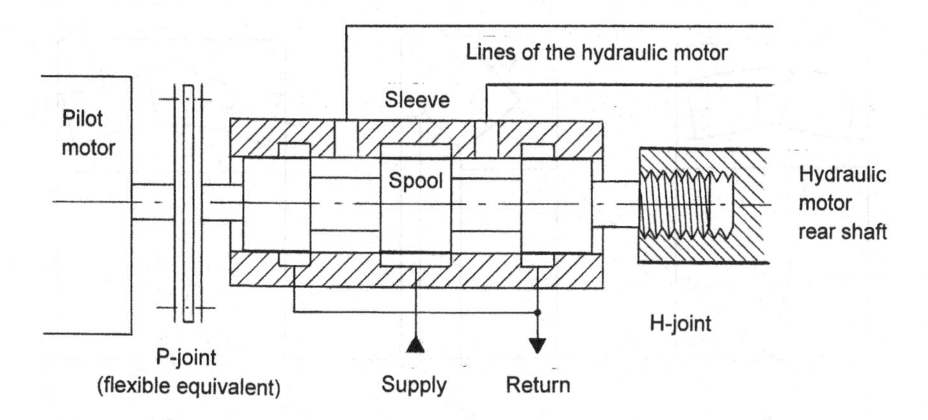

Fig. 8.6 Electrohydraulic pulse motor schematic

ingenious idea concealed therein was to mechanically interpose a servovalve spool of the described type between the pilot motor shaft and the hydraulic motor rear-extending shaft such as to assume a discriminator (comparator) function; sensing the difference in angular positions of the two motors and transforming that difference into axial displacement of itself to open the control edges by a proportional amount to drive the hydraulic motor in a *closing* sense. This is achieved by connecting the spool to the pilot motor shaft though a P-joint (prismatic; plunging, or an equivalent) and to the hydraulic motor shaft through an H-joint (helical; screw–nut) as shown schematically in Fig. 8.6. The servovalve spool is thus made to fulfill two functional requirements simultaneously:

1. Mechanical comparator function.
2. Control edge opening and fluid routing/throttling function; the servovalve function proper.

The fact that had never been mentioned is that the spool will be rotating inside its sleeve at the same rate of the motors, typically up to 1,000 rpm and faster. Hydraulic servovalves are not designed for their spools to rotate and suffer unnecessary rubbing inside their sleeves, leading to a fast rate of deterioration of the ultra-high precision interface. This is an example of improper association of functions; violating the design principle of separate allocation of functions. However, technical interest in such electrohydraulic pulse motors seems to have declined even before any solution to the problem was suggested; to decouple the two functions.

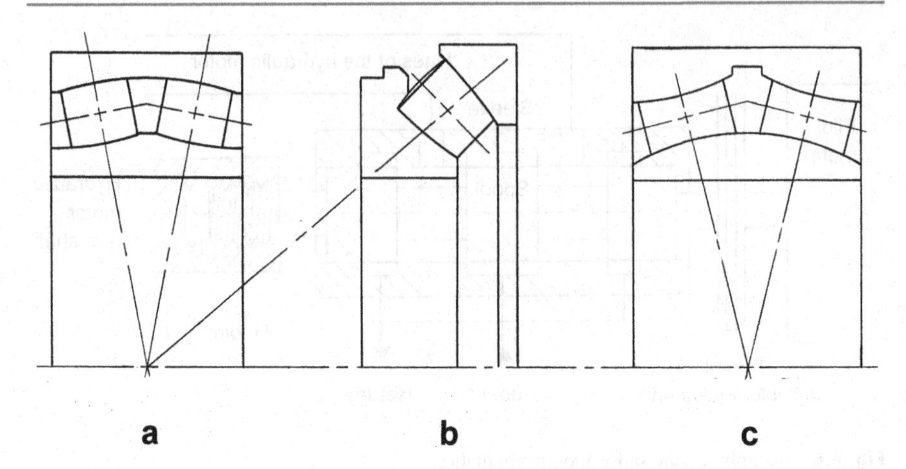

Fig. 8.7 Configurations of self-aligning roller bearings, **a** spherical roller bearing, **b** spherical roller thrust bearing, **c** spherical concave-roller bearing

8.4.2 Improper Association of Functions 2

One method for accommodating angular misalignment between two machine parts in relative rotation is to use self-aligning roller bearings, of which the three configurations thereof are shown schematically in Fig. 8.7. They consist of one or two rows of rollers brought to circular-arc line contact with a spherical race on one of the rings, and along a diametrically opposite generatrix with a ring-toroidal race on the other. It is the spherical feature that accommodates misalignment, generally of up to 1.5°, but only quasi-statically. Thus, these bearings embody the concept of associating or combining two functions in one unit; rotation and tilt. However, this exposes the designer to two potential pitfalls:

1. Correct fulfillment of the misalignment-accommodating function is conditional upon the rotating-bearing function being operative; otherwise the loaded rollers will rub on the spherical race in lengthwise direction and the surfaces will be damaged.
2. Since the shaft is usually fitted into the inner race of a spherical roller bearing (Fig. 8.7a) or a spherical roller thrust bearing (Fig. 8.7b), with the outer race fitted inside a hub or housing, correct rolling is obtained only when the misaligned shaft rotates; the rollers will be rolling on a slanted spherical segment of the inside spherical race. But if the misaligned hub was the rotating element on a stationary shaft, then the rollers will be wobbling; oscillating their track on the spherical race at too fast a rate. The bearing is said to be *abused*, and the wear rate will be much faster than anticipated.

Solution to the latter problem (of a misaligned rotating hub) is by using spherical concave- roller bearings (Fig. 8.7c). In these bearings the spherical feature is on the inner race. The concept of spherical concave-roller bearings is old, but they are now available as standard products from a few leading manufacturers.

8.5 The Pitfall of Conditional Functionality

Reliable execution of an allocated function should not be conditional upon uncertain variables to remain *arbitrarily small* or within tight limits, or upon another function being active or deactivated. Otherwise the mechanical assembly would inevitably fail. Therefore,

1. Components that associate or integrate two or more functional requirements should be carefully assessed for functionality, or else the design principle of proper allocation of functions should be respected. The case discussed in Sect. 8.4.2 is an example.
2. The commanded instantaneous position of an element in equilibrium under a system of forces should only negligibly be affected by friction or other uncertain variables, thence robust equilibrium.

References

Arafa HA (2006) Mechanical design pitfalls. Proc I Mech E Part C J Mech Eng Sci 220(6):887–899. https://doi.org/10.1243/09544062JMES185

Dudley DW (1962) Gear arrangements. In: Dudley DW (ed) Gear handbook. McGraw-Hill, New York, pp 3–1–3–44

Pahl G (1973) Prinzip der Aufgabenteilung. Konstruktion 25(5):191–196

Pahl G, Beitz W (1984) Engineering design. The Design Council, Berlin, London and Springer

Thoma H, Molly H (1939) Hydraulic device. US Patent 2,177,613, 24 October 1939

Gearing Design for High Power Density

9

High power density is one of the major attributes of gearing systems. It could primarily be achieved by adopting designs that feature multiple power paths, which make some of the gears multi-mesh gears, in particular the larger and heavier ones, and some of the pinions double-mesh ones. Since not all double-mesh pinions are in a driving or a driven situation, careful examination of the correct disposition of the backlashes for equal load sharing in all of the gear meshes will be presented with the purpose of avoiding running into design pitfalls regarding durability. A further method of increasing the power density using gears with asymmetric teeth will also be discussed. The application examples used in the context of these issues pertain to wind-turbine gearing as well as rotorcraft and turbofan transmission gearing.

9.1 Performance Density Criteria

Power Gearboxes

Speed reducing or increasing gearboxes—intended for one and the same generic application area—could be compared on the basis of their low-speed-end torque density. Despite that the volume and mass of the lower-speed components predominantly define the total mass (including the casing) these gearboxes come in a wide range of transmission ratios, such that *some* contribution of that ratio should be included. This means departure from the single-parameter specification; two parameters, properly merged into an empirical expression, are deemed to give a more closely accurate comparative assessment. Performance density will then be replaced by an empirical figure of merit (abbreviated FOM) of the gearbox design. And this is more meaningful when applied to a sufficiently large population of functionally identical/similar products. Examples include speed-reducing gearboxes for rotorcraft and fixed-wing turboprops, or speed increasing gearboxes for wind-turbine generators.

H. A. Arafa, *Design for Durability and Performance Density*,
https://doi.org/10.1007/978-3-030-56816-0_9

Aerospace Gearboxes

Brown et al. (2005) devised an empirical parameter called the power–speed index to include an effect of the reduction ratio (i). Plotted versus the gearbox weight on log-log scales, proportionality was revealed and the position of the best-fit straight line was indicative of the *modernity* of the design and technology; the technology trends of the years 1980 and 2000 were identified and a third futuristic trend line was drafted, each with its proportionality factor. The diagram includes data of more than 50 gearboxes of both helicopters and turboprop aircraft; both being intended for the same field of application. However, the power–speed index was formulated in such a way as to give an impression of it being a three-variable figure: $(\text{hp})^{0.76} \times (\text{engine rpm})^{0.13}/(\text{rotor rpm})^{0.89}$, whereas it should rather read $T^{0.76} \times i^{0.13}$, where T is the output shaft torque. Put in SI units, dividing by the gearbox mass, and including the given and anticipated values of the proportionality factor mentioned in the above reference, then

$$\text{FOM} = T^{0.76} \times i^{0.13}/m.$$

This empirical FOM (not dimensionless) should amount to 16, 20, and 25 approximately for the technologies of the years 1980, 2000, and the future ambitions, respectively. The expression explicitly shows that the output shaft torque is weighted by its 0.76 exponent and the reduction ratio by its 0.13 exponent, only, doubling the result as i increases from 1 to 207. This relationship is graphically plotted in Fig. 9.1 in the range of data covered by Brown et al. (2005).

Example A recently introduced heavy transport aircraft is powered by four turboprops of which the reduction gearbox could be assumed to have the following data:

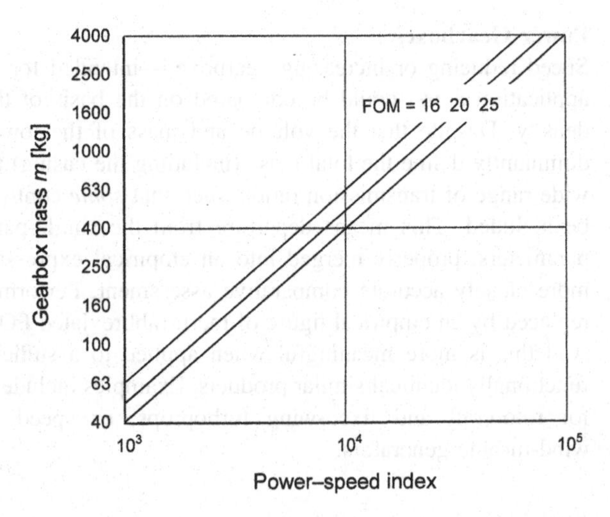

Fig. 9.1 Mass versus the power–speed index of aerospace gearboxes, hence the figure of merit (FOM) for different-modernity technologies

Maximum power = 8,100 kW
Maximum output speed = 850 rpm
Reduction ratio = 9.5
Weight = 460 kg
Therefore, the maximum output torque $T = 91{,}000$ N.m, the torque density $\approx$ 200 N.m/kg and the FOM $\approx$ 17.1, indicating a somewhat conservative design.

9.2 Split Power Paths

It has long been recognized that utilizing multi-mesh gears increases the power that could be handled by the individual gear(s), resulting in higher power density of the gear set. This becomes more pronounced as the larger and heavier gears, such as bull gears or internal gears, are designed as multi-mesh ones. This is achieved by gearing designs in which the input power is split at the input terminal into multiple power paths then recombined at the output terminal. These designs are referred to as power-split (or split power path) gearing, or split-torque transmissions. The simple planetary gear set is a known example. When higher transmission ratios than achievable with a single planetary gear set are required, then two such sets could be arranged in series. They should be of an appropriate size ratio for the same throughput power to be handled by each set at their respective speeds, hence no further gain in the power density. With an aim at increasing the torque density, White (2003) derived *the* sixteen possible versions of two interconnected planetary gear sets that do not conceal power recirculation, by virtue of power splitting then recombining, hence are expected to be of high energy efficiency. However, the element connections between the two sets are mostly intertwingled, except for one of these versions which could achieve a transmission ratio of a little above 25 with the gear proportions chosen close to practical limits, as shown schematically in Fig. 9.2. A more recent reinvention of this gearing is due to Berger et al. (2008), suggested as a wind-turbine transmission. Input power is split between the ring gears of a planetary set and a star gear set, to be followed by a simple step-up gear pair to achieve the ratios required for the intended application. Analysis of the relative values of forces and velocities within the system with the shown proportions results in power-splitting ratios 22% through the first planetary stage and 78% through the second-star set. Though unequal, this power splitting could contribute to increasing the torque density of the system, especially that the power recombining function is carried out within the first stage, requiring no additional means.

The above feature is to be compared with an alternative system of two planetary gear sets in a power-splitting configuration, for which the interconnection and the power recombination has to be accomplished in a third planetary gear set (in its two-DOF version; without grounded elements), as shown schematically in Fig. 9.3. This feature, sometimes referred to as a *differential*, implies some additional weight of the system. However, the proportions give better values of a transmission ratio of about 35 and power-splitting ratios of 68% and 32% in the first and second stages,

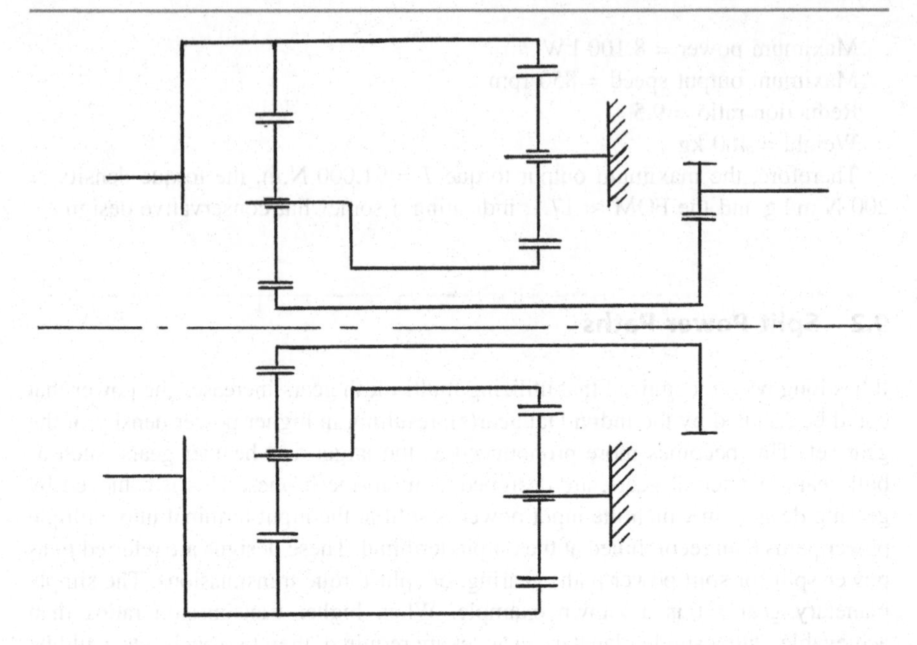

Fig. 9.2 Two interconnected planetary gear sets for power splitting then recombining, which do not conceal power recirculation

respectively. This gearing too should be followed by a simple step-up gear pair to achieve the ratios typical of wind-turbine transmissions, but also to have an open-center design which allows access to the rotor hub by mechanical or electrical means.

According to the latest report on this configuration of gearing by Strasser et al. (2018), it was proven in the field since about the year 2003. A fully detailed sectional drawing of this gearing with the same proportions is found elsewhere (Kejun et al. 2008). There are a number of other design versions and inversions of such three-set planetary gearing intended for the same application area, which are enumerated by Beck et al. (2014).

9.2.1 Double-Mesh Pinions, Disposition of the Backlash

Not all multiple meshing situations are intended for splitting the driving power. Therefore, it is essential to sort out gear/pinion multiple meshing cases in general, as to be in a driving and/or driven mode; power splitting or recombining, with essentially equal tooth loads, as well as to being in need of backlash adjustment or not, for correctly fulfilling is functional requirement.

A pinion meshing with two gears, not necessarily on diametrically opposite sides, could be envisaged to fulfill either one of two functional requirements:

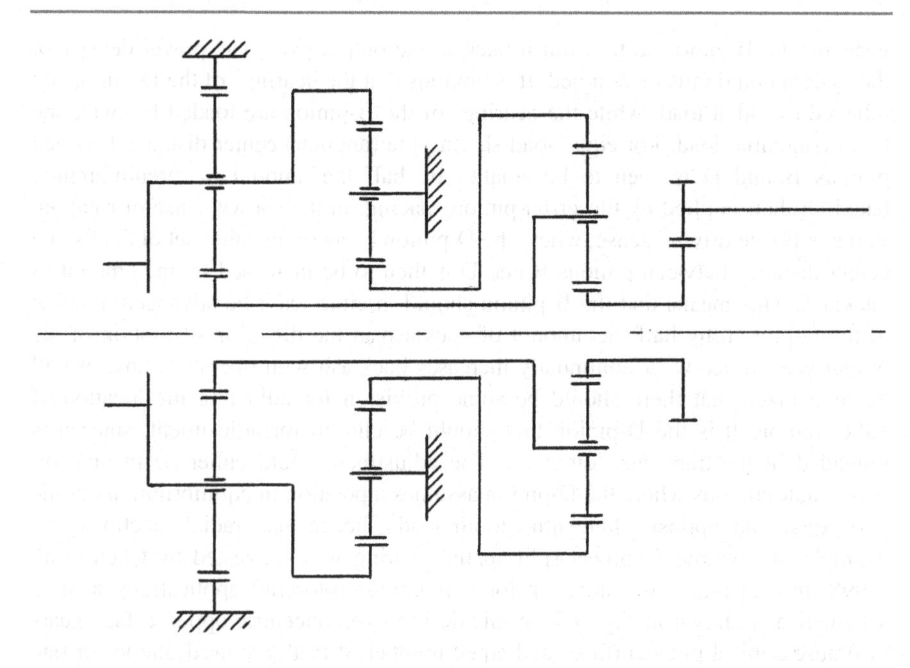

Fig. 9.3 System of two planetary gear sets in a power-splitting configuration, with a third set for power recombining

1. B-pinion; a Balance pinion (idler; load-transfer idler); a torque-free element that receives power from one of the gears and transmits the same power to the other gear. The two tooth loads being essentially equal.
2. D-pinion; a Dual Drive/Driven pinion, to transmit a torque in an equal power branching or recombining mode, for almost doubling the power density.
3. C-pinion; a power recirculating pinion; a combination of B- and D-pinions, where the two tooth loads are substantially different (see Sect. 3.1.1 and Fig. 3.3).

9.2.2 Asymmetric-Power-Split Spatial Gearing

Figure 9.4 is a partial section of two coaxial opposed face gears, developed into a plane, showing one input driving pinion (D) and one balance pinion (B). The section is through a cylindrical curtain at the middle of face width; to reveal the average pressure angle of the face gears. The (equal) backlashes are depicted the correct way; in the same rotational direction on D and oppositely on B. This is part of a so-called asymmetric-power-split gearing. The asymmetry stems from the differently routed power paths: the D-pinion transmits half an input power directly to the upper output gear and the other half power to the lower load-transfer (idler) gear, for the B-pinion to transmit it back to the output gear. The power density of the system could thus be doubled. It is obvious that the bearings of the D-pinion are relieved of radial load, while the bearings of the B-pinion are loaded by twice the tooth tangential load. For equal load sharing the (angular) center distance between pinions B and D is seen to be smaller by half the amount of circumferential backlash than implied by the *gross* pinion spacing, in the shown sense of rotation. In the opposite driving sense, where the D-pinion bears on its other set of flanks, the center distance between pinions B and D is then to be increased by the amount of backlash. This means that the B-pinion should in either case be advanced, relative to the D-pinion, by half the amount of backlash in the direction of motion of the output gear. Since wear continually increases backlash with operating time, it will be recognized that there should be some provision for adjusting the location of either pinion. It is the D-pinion that should be chosen for adjustment, since it is unloaded in the transverse direction. The adjustment could either be manual or, better, autonomous where the D-pinion assumes a position in equilibrium under the two equal and opposite tangential tooth loads, hence also radial reactions. An example of asymmetric-power-split spatial gearing was suggested by Chen et al. (1998) in a face-gear transmission for twin-engine rotorcraft applications such as schematically shown in Fig. 9.5. In this design two concentric opposed face gears of obtuse conical pitch surfaces and equal number of teeth are used; the lower one being the load-transfer gear and the upper one the power recombining, output gear. Two *diametrically opposite* spur B-pinions are meshed between the face gears, rigidly supported in the housing by being straddle-mounted in their bearings, and two diametrically opposite input spur D-pinions are kept *floating* by being the end regions of rather long, compliant cantilever shafts; the two input shafts. This scheme is a duplication of the one shown in Fig. 9.4. The following conditions for proper functioning apply.

1. For the two B-pinions to be equally loaded the four backlashes between them and the face gears should be exactly equal, other mix and match situations being excluded.

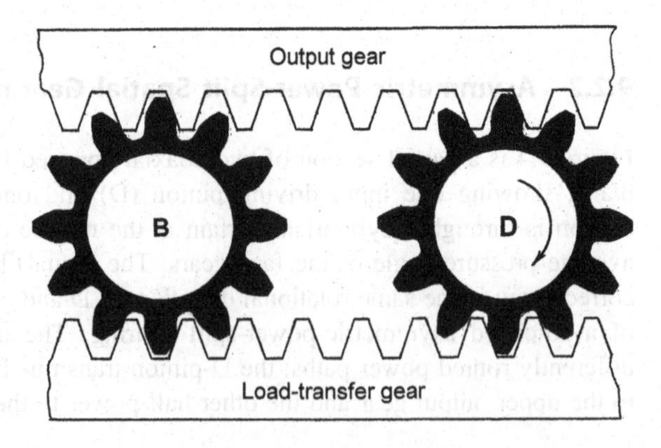

Fig. 9.4 Double-mesh pinions as an input drive pinion (D) with same-direction backlashes and a balance pinion (B) with opposite backlashes

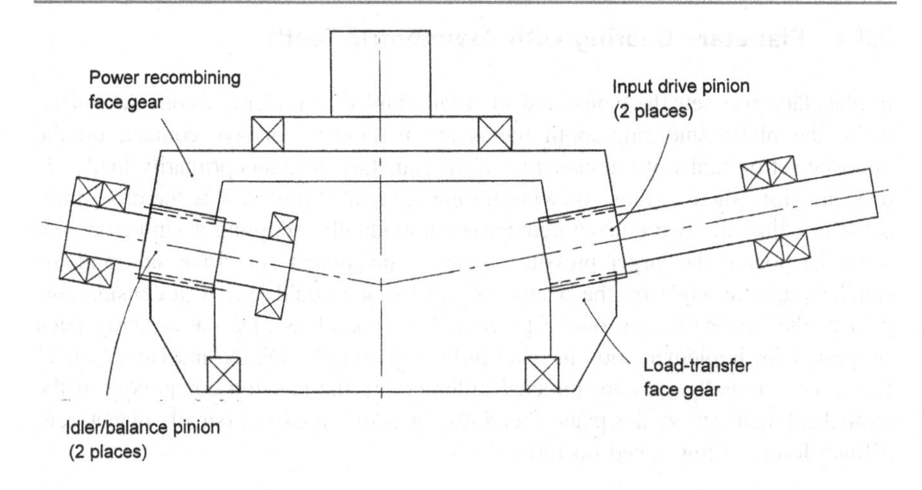

Fig. 9.5 Schematic of a split-torque face-gear transmission with two inputs

2. For the two D-pinion shafts to remain straight the diameter joining the centers of the two B-pinions should be angularly advanced in the direction of rotation of the power recombining, output face gear by half the amount of circumferential backlash. Otherwise, bent input shafts will result in unequal power splitting.

9.3 Asymmetric Gear Teeth

The load-carrying capacity of gears could be improved by increasing the pressure angle. This increases the radii of curvature of the flanks, reducing the Hertzian contact stress, and enlarges the spread of the elasto-hydrodynamic film of lubricant. This tendency is found in the design of aerospace gearing; pressure angles of 25° to 28° being commonplace. But the contact ratio will be compromised.

Conventional, symmetrical gear teeth have the advantage of rational machining by well-established processes and readily available tooling. In case of unidirectional loading there also exists the possibility of the individual gear to be inverted for extending the useful life on the opposite, unused set of tooth flanks (the coast side). But increasing the pressure angle of symmetrical gear teeth could lead to unacceptably small top land; even tooth pointing. Therefore, asymmetric teeth (buttress teeth) have often been suggested for unidirectionally loaded gear sets. These are teeth that have on their drive side a larger pressure angle, of involute profiles drawn to a smaller base circle, and vice versa on their coast side. The same is done on the mating gear. Neither the top land nor the root fillets will be at a disadvantage, and the tooth cross-section will remain well balanced.

9.3.1 Planetary Gearing with Asymmetric Teeth

In planetary gear sets the planet and sun tooth flanks are in convex-convex contact, while the planet and ring tooth flanks are in convex-concave contact, on the opposite set of flanks of the planet teeth. In planetary gear sets primarily loaded in one direction, such as those of wind-turbine generator drives, it is therefore indicated to adjust the two contact geometries individually, adopting asymmetric gear teeth. Disposing the larger pressure angle on the planet–sun drive side and the smaller pressure angle on the planet–ring drive side could lead to a considerable gain in the power density of the gearing, Fig. 9.6a. This idea has recently been suggested for implementation in wind-turbine gearing by Dinter and Hess (2012). However, a radially outward directed difference in the reaction component of the tooth load will act on the planet bearings, in addition to the (small) planet centrifugal force, in low-speed operation.

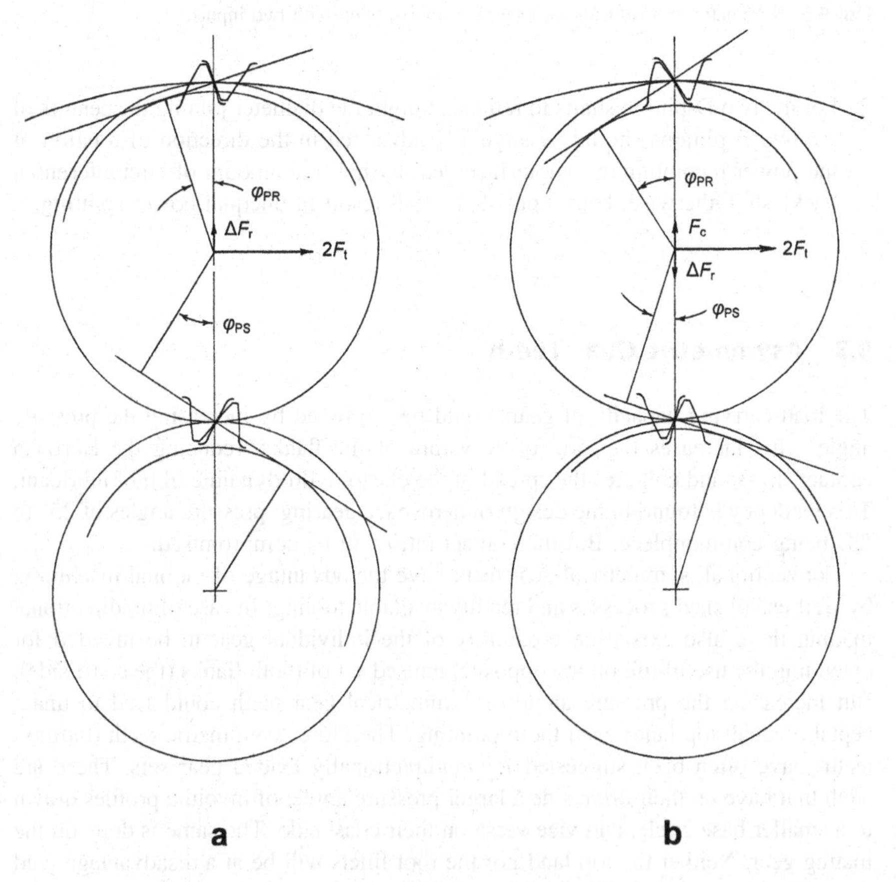

Fig. 9.6 Planetary gearing with asymmetric teeth for higher power density, **a** for low speeds, **b** for high-speed operation

In high-speed planetary gearing the planet bearings are subjected to a substantial centrifugal force that, in combination with (twice) the tangential tooth load, could markedly impair their life expectancy. Such situations require a trade-off between the durability of the tooth flanks and that of the planet bearings, so that the inverse disposition of the pressure angles would even be suggested; the larger pressure angle on the planet–ring drive side flanks and the smaller one on the planet–sun drive side flanks, Fig. 9.6b. The resulting difference in the radial-reaction components of the tooth loads could largely counteract the planet centrifugal force. The latter suggestion was made by Venter and Krüger (2017) for implementation in aircraft turbofan gearboxes. The result of this action could be maintaining the good power density and having a consistent durability of the highly loaded interfaces. However, the tooth flanks with the smaller pressure angle could be considered at a disadvantage; the planets having no all-time *coast sides* as such. Both examples in Fig. 9.6 adopt pressure angles of 18° and 32° to be used conversely for the two planet meshes, to result in a ratio of the net radial reaction to the tangential tooth load in both cases of $\Delta F_r/F_t = \tan 32° - \tan 18° = 0.3$.

9.4 Compound Planetary Gearing

The simple planetary gearing is by its very nature of a split power path configuration, hence known to be of reasonably high power density. It is also of a balanced radial-reaction design and offers higher transmission ratios than non-planetary gearing. Still higher transmission ratios (of up to 20) could be obtained from compound planetary gearing of the same radial dimensions as simple planetary gearing. Figure 9.7 is a schematic of the basic compound planetary gearing that consists of one ring-less set and one sunless set compounded together through a twofold connection: (1) each two planets are made in one piece (referred to as a cluster or compound planet) and (2) they are supported on a common carrier. However, accurate relative angle phasing of each two planets in a cluster is an essential condition for assembly and, more importantly, equal load sharing. This is one reason why automotive automatic transmissions never include compound planetary sets.

Since helical gears are superior to spur gears regarding running quietness, running-in characteristics, and strength, compound planetary gearing is often designed with helical gears. Figure 9.8 shows a compound planet, the two gears of which are of different diameters, typically of the same helix hand but in the particular case of having the same lead (L), with their helical tooth traces on the pitch cylinders. The tangential tooth loads F_{t1} and F_{t2} are applied in diametrically opposite planes; from the sun and the ring gears. The compound planet, being a torque-free element, then $d_1F_{t1} = d_2F_{t2}$. The lead angles are given by $\tan \lambda_1 = L/(\pi d_1)$ and $\tan \lambda_2 = L/(\pi d_2)$. The axial reaction components are given by $F_{a1} = F_{t1}/\tan \lambda_1$ and $F_{a2} = F_{t2}/\tan \lambda_2$, which will then be equal and opposite. Therefore, compound planets of the same lead exert no axial reaction on the carrier; they

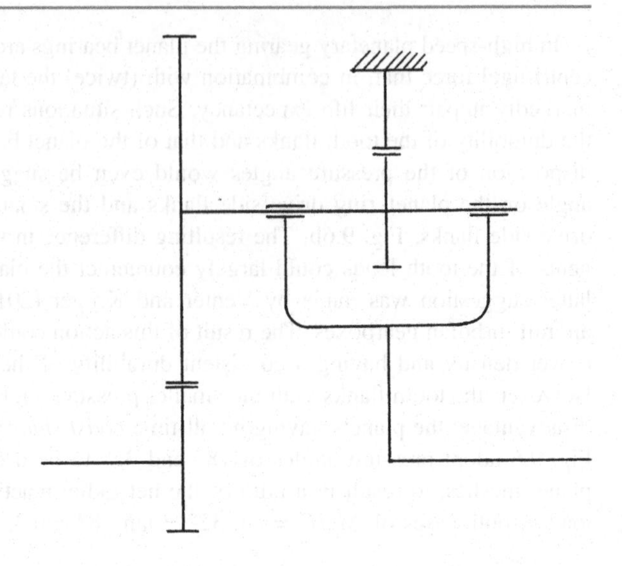

Fig. 9.7 Basic compound planetary gearing with one sun and one fixed ring gear

would assume whatever axial location without altering the system phasing, as if they were spur gears. Therefore, they require precise relative timing to be assembled into the system and to provide equal load sharing. It is also known that when the two gears of a compound planet are of different leads they will react on the carrier in a certain direction and with a certain axial force that could easily be calculated. This very feature is made use of in equalizing the load sharing (without resorting to angle phasing) by axially adjusting each planet cluster, maybe with some selective assembly to replace the accurate timing.

9.4.1 Compound Planetary Gearing with Asymmetric Teeth

Compound planetary gearing in biplanar design such as shown in Fig. 9.7 offers the possibility of making all the tooth drive sides of the higher pressure angle. This is by virtue of the two planets (in a cluster) each having its own drive and coast sides, and the teeth of the two planets could be made oppositely buttressed. In their endeavor to devise "improved gear teeth," Binney et al. (2019) suggested a similar system for rotorcraft transmissions with even larger pressure angles of 20° and 36° for the coast side and the drive side, respectively, of the (heavily loaded) planet–ring mesh in one set as well as for the planet–sun mesh in the other set. The problem with compound planetary gearing—that it could accommodate a rather limited number of the smaller planets engaging the ring gear—was resolved by using a staggered layout of overlapping large planets to pack a larger number of planet clusters, hence to further increase the torque density. However, the necessity of accurate relative phasing of all the twin planets in compound planetary gearing, or else their adjustment (when of different lead) should not be overlooked.

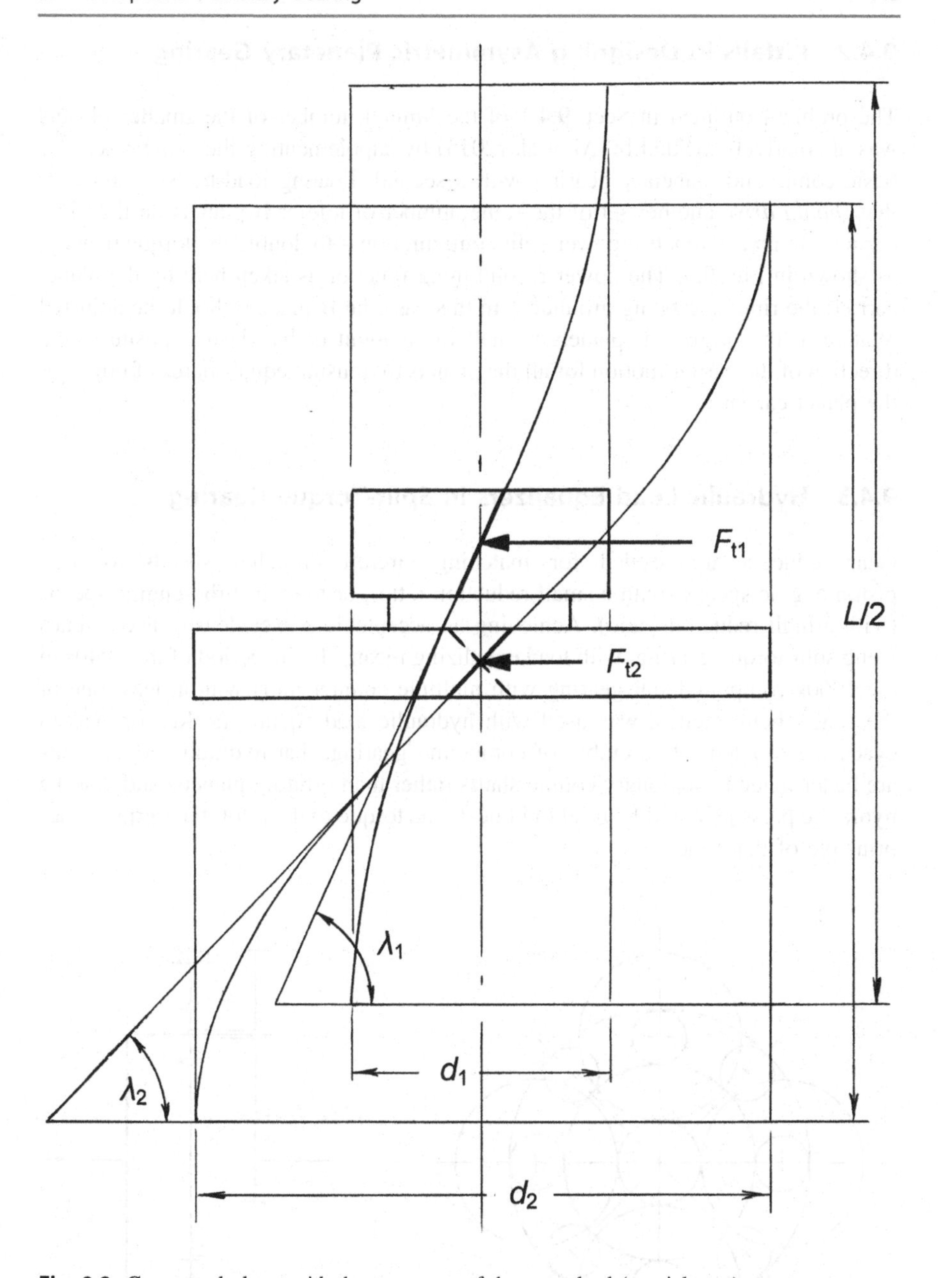

Fig. 9.8 Compound planet with the two gears of the same lead (special case)

9.4.2 Pitfalls in Designing Asymmetric Planetary Gearing

The problem outlined in Sect. 9.4.1 of the limited number of the smaller planets was alternatively tackled by Ai et al. (2015) by supplementing the sunless set of a basic compound planetary gearing with a second, floating load-transfer sun gear that *should drive* another set of the same number of idle or B-planets on the same carrier—in an asymmetric-power-split configuration—to double the torque density, as shown in Fig. 9.9. The power recombining function is taken here by the planet carrier; the ring gear being grounded. In this case the B-planets should be adjusted relative to the original D-planets by half the amount of backlash opposite to the direction of the carrier motion for all the planets to transmit equal shares of power to the planet carrier.

9.4.3 Hydraulic Load Equalizers in Split-Torque Gearing

Gear reducers are needed for matching aircraft propeller speeds to their piston-engine speeds (with a small reduction ratio), or to their turbo-engine speeds (with a high reduction ratio). Achieving an acceptable power density necessitates using split-torque gearing, with load equalizing means. In the period of the 1940s to the 1960s compound star gearing with multiple countershafts with at least one of the stages being helical was used with hydraulic load equalizers. Reason was to evade the problem of assembly of compound gearing, that hydraulic connections are better made to stationary countershafts rather than orbiting planets, and that the hydraulic pressure could be used to indicate the torque to the pilot; the torque-meter principle of that time.

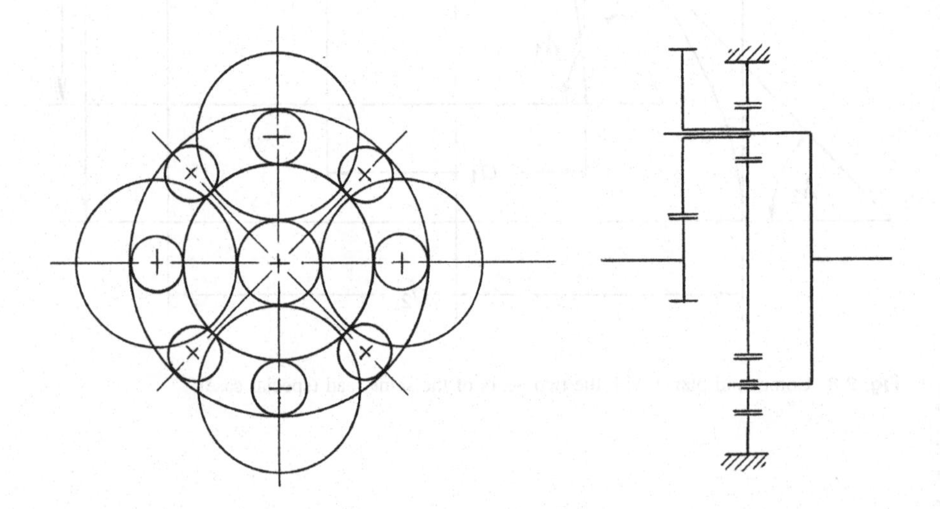

Fig. 9.9 Asymmetric-power-split compound planetary gearing with fixed ring gear

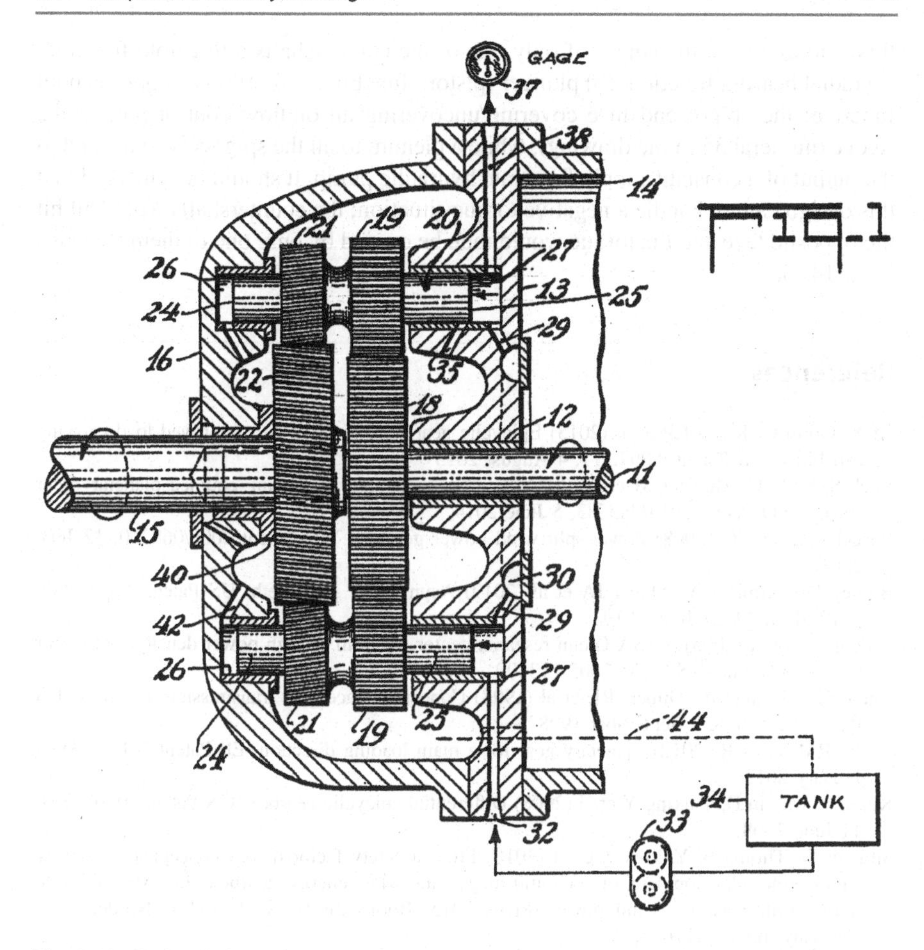

Fig. 9.10 Hydraulic load equalizer for propeller aircraft; the torque-meter principle. Reproduced from Taylor (1945), public domain

The earliest document on hydraulic load equalizers is the patent by Taylor (1945), the original drawing figure of which will be used here to explain the underlying principle, since it is the mostly generic of all the suggestions that followed, Fig. 9.10. The input stage features spur gears and the output stage helical gears, viz., a compound gearing with the planets of different lead. For equal load sharing each of the countershafts should carry the same torque, hence be subjected to the same axial reaction force. Since this condition could only be satisfied when each countershaft assumes its own axial location that brings the drive flanks of both the spur and the helical pinions to bear, then the axial loading should be made insensitive to the axial location of each countershaft. This is achieved by subjecting the end faces of all the countershafts to the same fluid pressure, which should also be maintained proportional to the torque. A straightforward method of achieving this is to assign to the spigot of only one of the countershafts a threefold function: (a) radial bearing function, (b) plunger (piston) function, and (c) servo-control spool function; the spigot end face covering/uncovering an outflow control port in the sleeve (numeral 35 in the drawing), while a plenum to all the spigots is connected to the output of a constant supply flow-rate hydraulic pump. It should be remarked that this design cannot handle a negative torque situation; the countershafts would all hit the opposite face and the torque would then be carried by only one of them (see also Sect. 14.3).

References

Ai X, Orkin C, Kruse RP et al (2015) Epicyclic gear transmission with improved load carrying capability. US Patent 9,097,317, 4 August 2015

Beck S, Sibla C, Münster M et al (2014) Transmission unit comprising coupled planetary gear stages. WO Patent 2014/082783, 5 June 2014

Berger G, Bauer G (2008) Power-split wind power gearbox. WO Patent 2008/068260, 12 June 2008

Binney DA, Gmirya Y, Mucci JA et al (2019) Asymmetric gear teeth. US Patent Application 2019/0195329, 27 June 2019

Brown GV et al (2005) NASA Glenn research center program in high power density motors for aeropropulsion. NASA TM 2005-213800

Chen YJD, Heath GF, Gilbert RE et al (1998) Concentric face gear transmission assembly. US Patent 5,802,918, 8 September 1998

Dinter RM, Hess R (2012) Planetary gear for a main loading direction. EP Patent 2 402 631, 4 January 2012

Kejun Y, Muxin D, Caixing Y et al (2008) Differential epicyclic gearbox. CN Patent 201071904, 11 June 2008

Strasser D, Thoma F, Yüksek S et al (2018) From a safety factor driven concept to reliability engineering: development of an multi-mega-watt wind energy gearbox. In: Abel D et al (eds) Conference for wind power drives 2015. Books on Demand GmbH, Norderstaedt, Germany, pp 153–176

Taylor ES (1945) Gear system. US Patent 2,386,367, 9 October 1945

Venter G, Krüger D (2017) Planetengetriebevorrichtung und Strahltriebwerk mit einer Planetengetriebevorrichtung. DE Patent 10 2015 122 813, 29 June 2017

White G (2003) Derivation of high efficiency two-stage epicyclic gears. Mech Mach Theory 38:149–159. https://doi.org/10.1016/S0094-114X(02)00093-9

High Power/Torque-Density Devices

10

Mechanical power transmission and handling devices are evaluated according to their power density, or else their torque density if it is not the power that is the issue of primary concern. The combination of high power or torque density with good energy efficiency may collectively be termed high-performance density. In continuation to the deliberations on gearing design for high power density in Chap. 9, various mechanical devices are presented here with the features and particulars that make them measure up to being of high-performance density. Due attention is given to the details of long high-speed shafting, full-complement differential gearing, planetary roller screws in their various configurations, constant-velocity shaft joints, multi-disc clutches, and helical-rotor gear pumps.

10.1 Long Shafting of High Power Density

Designed with an ample margin before elastic torsional instability (torsional buckling), a tubular shaft is known to have a much higher torque density than a solid one. The critical speed of a tubular shaft is also higher than that of a solid one designed for the same torque. Therefore, a tubular shaft has an appreciably higher power density than a solid shaft, in the range of elastically and dynamically stable operation.

To maintain the same attribute with long shafting the tubular shaft should be divided into a number of shorter sections of reasonable length, connected end-to-end by coupling means, each of which being supported in a radial bearing in a hanger bearing assembly, making a modular construction.

Typical application of multi-section tubular shafts is in the tail rotor drive shaft of helicopters and in the drive shafts to the forward and aft transmissions of tandem-rotor helicopters. An example sectional drawing of a hanger bearing assembly is shown in Fig. 10.1. The tubular shafts are provided with a face-toothed coupling flange at each end (with concave or convex tooth flanks), which engage

H. A. Arafa, *Design for Durability and Performance Density*,
https://doi.org/10.1007/978-3-030-56816-0_10

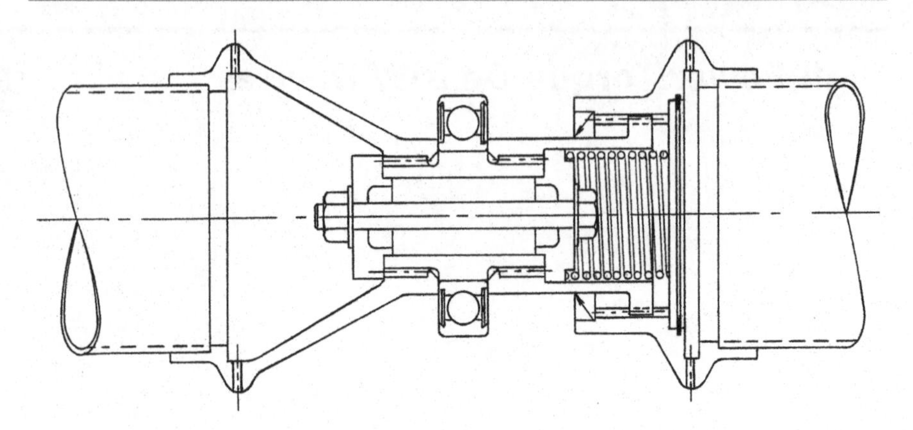

Fig. 10.1 Hanger bearing and coupling assembly in long multi-element shafts

complementary teeth in the flanges of the coupling hub, and each pair is secured by a U-shaped split clamping ring (not shown). The hanger bearing assembly consists further of a short hollow shaft, splined over both ends to engage the coupling hubs, and is supported in its middle in a sealed, low friction deep groove ball bearing. The bearing is supported through an elastomeric damper in the hanger (not shown). One of the coupling halves includes a spring-loaded splined joint to accommodate axial float, and to allow taking out any of the hanger bearing assemblies for replacement.

10.2 Differential Gearing of Higher Torque Density

The torque density of differential gearing increases with the number of planets disposed between the two output side gears. Highest torque density is therefore achieved with the so-called full-complement, parallel-axis differentials, usually made all-helical. Two opposite-hand sets of *long* planets are provided, for each set to engage one of the side gears and for each planet in a set to engage the two adjacent planets in the other set.

A schematic exploded view of a full-complement, all-helical differential after is shown in Fig. 10.2. The original idea is after Quaife (1985). Each planet (or pinion) is supported in a partial-cylindrical cavity in the differential casing. The deeper half-lengths of these blind cavities starting at one end of the differential casing intersect those from the opposite end to make all the pinions intermesh in the middle zone. The flat bottoms of the cavities are just short of the opposite side gear. The middle zone contains a means for axially preloading the side gears by Belleville springs (not shown). The frictional resistance to rotation of the planets and the side gears when the vehicle negotiates a turn makes this device a so-called limited-slip differential. Certain strict conditions pertain to the assembly of such full-complement planets, and these are given hereinafter.

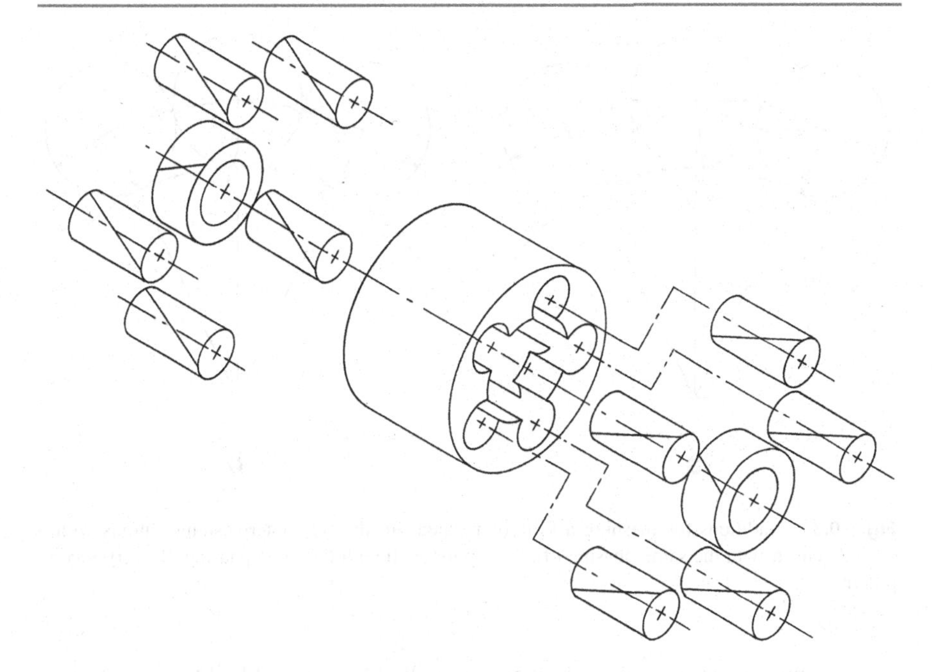

Fig. 10.2 Full-complement, all-helical differential in an exploded view

There also exist a number of parallel-axis, all-helical differentials with two or three independent intermeshing planet pairs. These will assemble unconditionally, but they are of less torque density.

10.2.1 Assembly of Full-Complement Planet Pairs

All-helical differentials having a full complement of intermeshing planet pairs form an interlocked train of gearing as shown in Fig. 10.3; every other planet engaging the other side gear. To arrive at the assembly conditions of these differentials first consider an all-spur version; the results being applicable to all-helical sets as well. Since, with a fixed cage, the two side gears can rotate in opposite directions, the axial view will be considered while the side gear teeth coincide. The problem becomes one of determining the assembly conditions for a full complement of an even number of intermeshing spur pinions around one central gear, which form an immobile configuration.

In general, any two intermeshing pinions will engage a gear with the three centers on the vertices of a triangle, and the assembly will be locked. Assuming identical pinions, adding a third one will only be possible in axial direction and when the same orientation of the tooth–space meshes of the first two pinions (in one case, Fig. 10.3a) or the orientation of the first pinion (in a second case, Fig. 10.3b)

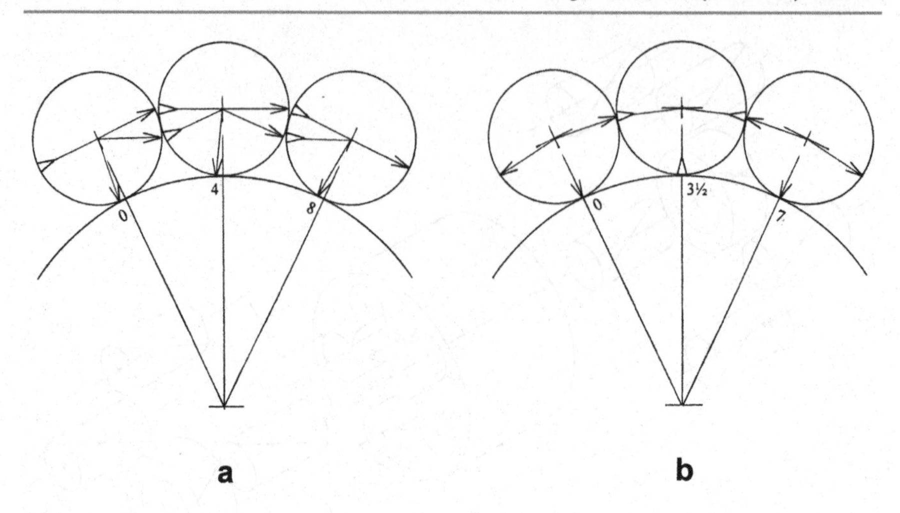

Fig. 10.3 Conditions for meshing a full complement of identical, intermeshing pinions with a central gear, **a** with the same phasing of all the pinions, **b** with the same phasing of every second pinion

is replicated. In other words adding pinions will only be possible if the system is so dimensioned as to accommodate a full complement thereof about the central gear. The following two conditions should thus simultaneously be satisfied.

1. The number of teeth on the central (sun or side) gear N_S should be divisible by the number of pinions q for identical phasing of all the pinions relative to the side gear, or by half the number of pinions for identical phasing of every second pinion (see also Sect. 10.2.2).
2. The pinion-to-pinion center distance c_{PP} and the pinion-to-sun center distance c_{PS} should exactly give the angular spacing commensurate with the number of pinions; $c_{PP}/2c_{PS} = \sin 180°/q$. This must not exactly equal the rational fraction $N_P/(N_P + N_S)$ because of the necessary introduction of different profile shifts to the side gear and/or the pinions to adjust the center distances.

Saari (1966) was first to derive assembly conditions in a somewhat different way (using the pitch circle radii, without referring to the profile shifts), which might cause some confusion. The full-complement differentials were also suggested in the same work.

10.2.2 Worked Examples

The all-helical differential with a full complement of planet pairs is typically designed with a number of planets $q = 10$, side gears of 15 teeth (divisible by $q/2$), and planets with 6 teeth. Following are two calculation examples with more planets, of larger numbers of teeth, for any different application, to show the two possible relative orientations of the tooth meshes. In Fig. 10.3, a pinion tooth is represented by an outward pointing V and a tooth space by an inward pointing one. The calculations are somewhat approximate because of the simplified formula for the center distance in presence of profile shift (x), but they emphasize the need for such. The profile shift is to be determined for the pinions (due to their smaller number of teeth); the side gears being left as standard. Therefore,

$$x_P = (N_S/2)/[(1/\sin 180°/q) - 1] - N_P/2$$

Example 1 $N_P = 15$; $q = 14$; $N_S = 56$; hence $x_P = +0.514$

In the reference position shown in Fig. 10.3a the phasing of all the planets relative to both side gears is the same. The successive planets mesh with the side gears every four pitches.

Example 2 $N_P = 13$; $q = 14$; $N_S = 49$; hence $x_P = +0.512$

In the reference position shown in Fig. 10.3b the phasing of every second planet relative to its side gear is the same. The successive planets mesh with the side gears every 3½ pitches.

10.3 Planetary Roller Screws

Planetary roller screws are a class of rotary-to-linear motion transformers of high energy efficiency and power density. The latter attribute is achieved by providing a multitude of rolling contacts of planetary rollers between the screw and nut well in excess of those in customary ball screws of similar dimensions, and by the contacts being short straight lines or else spots of a rather small curvature mismatch in all directions, contrary to the ball screws that provide a small curvature mismatch only in the axial plane. This latter fact makes the load carrying capacity of the contact spot much higher for the same Hertzian contact stress. Further virtues include elimination of the anomalous asymmetry typical of the ball screw loaded races and the non-existence of ball recirculating tubes, where failure is often triggered.

The introduction of the original type of planetary roller screws is attributed to Strandgren (1954), which configuration is being manufactured—much similar to the firstly suggested—by renowned companies until to date. Two other types of roller screws have evolved over the years, first with a purpose of making the positioning accuracy and repeatability insensitive to slip between the rolling elements, then with a purpose to simplify the design and to achieve highest possible power density. The main feature (planetary rollers) seems to group the different types in a collective, despite that the kinematics of the individual types is greatly different. The three types of commercially available planetary roller screws are described hereinafter, together with the kinematics thereof. Further design suggestions of less relevance are not considered.

The attribute of high power density is being used to describe planetary roller screws only since around the year 2010 by the manufacturing companies; that description was anyway not in common usage for long before. In addition to custom-made screws, manufacturers produce self-contained, limited-stroke actuators made up of a (servo) motor and some type of planetary roller screw. These units provide adequate lubrication and much better protection against ingress of contamination than the wiper seals on the screws themselves. They are referred to as servo screws, electric cylinders, or electromechanical cylinders. In the application areas where high loads are anticipated these actuators represent a viable alternative to hydraulic cylinders.

10.3.1 Threaded-Roller Screws

Nut Assembly The nut assembly of the firstly introduced type of planetary roller screws features a multi-start-threaded nut that meshes with a number of equally spaced, single-start threaded rollers. The nut and roller screw threads have the same axial pitch, helix hand, and lead angle, $\lambda_N = \lambda_S$, typically about arctan 0.1. Originally the nut thread had a V-shaped profile and the roller thread flanks were slightly convex, to reduce friction. But now, they are more correctly made with involute helicoid flanks. The nut assembly is intended for pure planetary motion of the rollers, just like in planetary helical gear sets. Therefore, the size ratio of the nut and rollers is expressed by the ratio of numbers of starts; an integer ratio which equals the number of starts in the nut N_N in case of single-start rollers.

To maintain pure planetary motion with such a small thread lead angle the rollers have spur-pinion teeth cut into both end regions on a pitch-circle diameter equal the mean thread diameter, which teeth engage a ring gear fixed to the nut at each end, Fig. 10.4. The gear ratio is thus equal N_N as well; an integer. Axial retention of the rollers is provided by the screw threads, while angular synchronization is maintained by the gear mesh, independently, and not requiring exact timing of the gear mesh with their roller thread start. This makes a so-called form closure; not depending on traction at the point-line contacts. The schematic drawing depicts the meshed ring gear and pinions, and the nut thread starts in an example with a ratio of 70/14 = 5 and with eight rollers. The nut assembly further includes two free guide rings to maintain equal spacing of the rollers. It thus ensures pure planetary motion of the rollers without any axial movement.

Lead Screw The lead screw has an N_S-start thread of involute helicoid flanks. Regardless of the hand of the helix the assembly condition of the system is identical to that of a planetary gear set: $(N_N + N_S)$ should be divisible by the number q of equally spaced rollers. Since the number of rollers is relatively large and the number of starts of either screw thread is rather small, the condition simplifies to $N_N + N_S = q$. In the present example N_S should equal 3. There are two cases to present in the following.

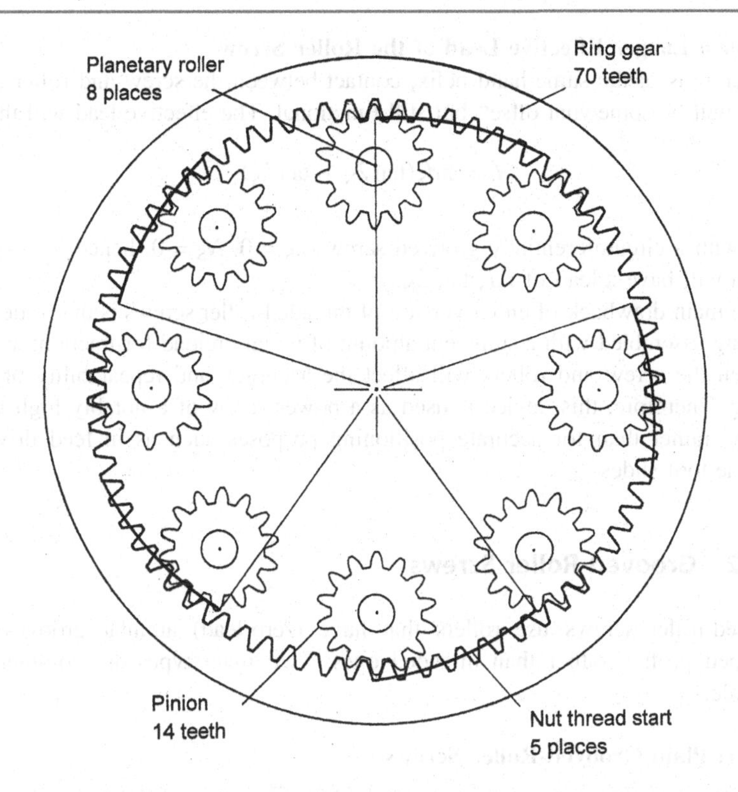

Fig. 10.4 Nut assembly of a threaded-roller screw with eight planetary rollers and five starts of the nut thread

(a) For a Small Effective Lead of the Roller Screw

The lead screw is of opposite-hand helix relative to the nut and rollers; the system would act exactly as a planetary gear set without any axial movement of the lead screw if the latter had the same lead angle as the rollers and nut $\lambda_N = \lambda_S$ (just like in the present example), hence it will not even be possible to introduce the lead screw by threading into the nut assembly in presence of the necessary radial preload. Changing q to 7 or 9 will change N_S to 2 or 4 and λ_S to smaller or larger values, respectively. The effective lead, without slip, would then be

$$L = \pi d_S(\tan \lambda_N - \tan \lambda_S)$$

where d_S is the effective diameter of the lead screw. This is a rather small lead (around $0.1d_S$), as a left-hand or a right-hand helical pair, respectively.

(b) **For a Larger Effective Lead of the Roller Screw**
The screw is of the same-hand helix, contact between the screw and roller thread flanks will be somewhat offset, but still functional. The effective lead will then be

$$L = \pi d_S(\tan \lambda_N + \tan \lambda_S)$$

Even with a circumferentially grooved screw ($\lambda_S = 0$, $N_S = 0$, hence $q = N_N$) the system will have a lead of $\pi d_S \tan \lambda_N$.

The main drawback of either version of threaded-roller screws is that it depends on being assembled with a sufficient amount of radial preload for traction; any slip between the screw and rollers will affect the accuracy and repeatability of positioning. Therefore, this device is used as a power screw of a notably high power density, rather than for accurate positioning purposes such as in feed drives of machine tool slides.

10.3.2 Grooved-Roller Screws

Grooved-roller screws use rollers that have (zero-lead) annular grooves of a V-shaped profile, rather than threaded ones. Two main types are commercially available.

Type 1: Plain Grooved-Roller Screws

This type was first suggested by Saari (1986). The (non-planetary) rollers have grooves of a flank angle between 30° and 45°, and their equal spacing and parallelism are maintained by their pintles guided in bearings in two fixed end caps of the housing as shown in Fig. 10.5. The nut is a rotating load-transfer member with internal annular grooves that engage the rollers, and is axially restrained to rotate between two thrust bearings in the housing. It is centered on the rollers, which should not be less than three in number.

The screw is a multi-start thread of involute helicoidal flanks, the number of starts being equal to the number of rollers, by necessity; typically five or six, sometimes eight. The screw looks therefore like a *fine* thread to accommodate the large number of starts on a limited diameter. The effective lead equals the lead of the screw, hence not smaller than three times the axial pitch. The main advantage of this system is that it is the one and only type where slip between any of the members does not affect the accuracy or repeatability of positioning. It is therefore not sensitive to the tightness of radial preload in assembly.

Grooved-roller screws should be of high durability because the contact between the involute helicoid flanks of the screw thread(s) and the obtuse conical flanks of the roller grooves is along a multitude of short straight lines, rather than on points. An involute helicoid is shown in Fig. 10.6 with its axis tilted by its base lead angle λ_b. In this particular projection one generatrix is oriented along the line of sight; the point

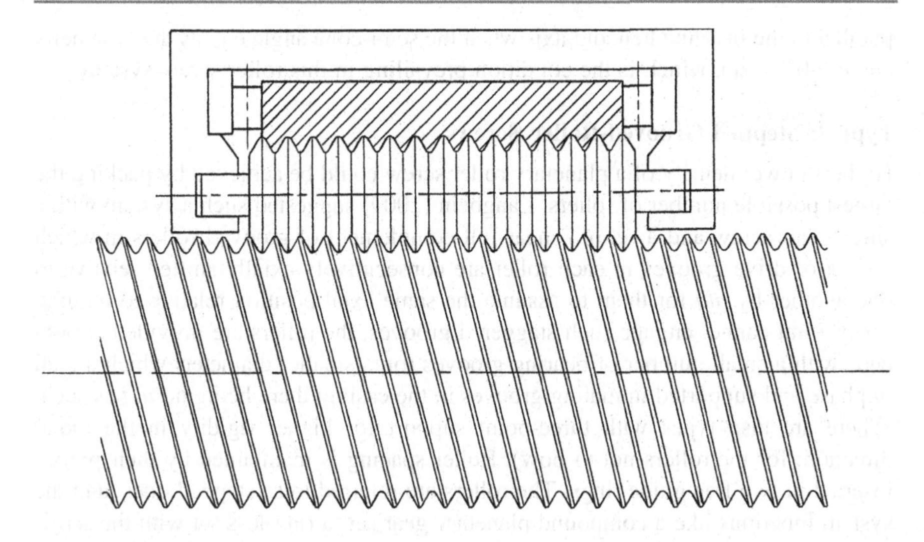

Fig. 10.5 Plain grooved-roller screw, showing one roller of a set of five; the screw has a five-start thread

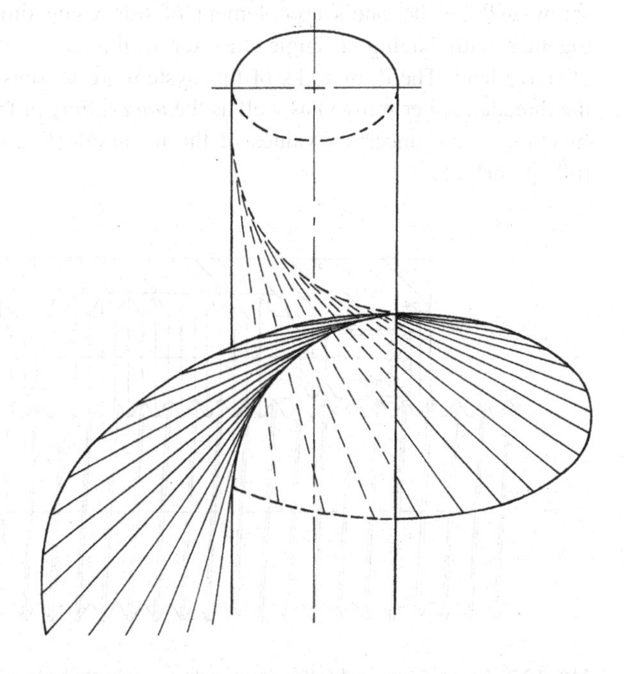

Fig. 10.6 Involute helicoid, projected with its axis tilted by the amount of the base lead angle of 40°. The inside, concave surface is seen to the right and the outside, convex surface to the left

of intersection of both helices on the perimeter and the base cylinder. It is evident that the outside (convex) helicoid surface can be in straight-line generatrix contact with any cone of an axis that is tangent to the base cylinder. The cone axis will be parallel to the involute helicoid axis when the semi-cone angle equals the base helix angle ($90° - \lambda_b$), which is the condition prevailing in this roller screw system.

Type 2: Stepped Grooved-Roller Screws

Highest power density of a planetary roller screw could be achieved by packing the largest possible number of rollers. Langbein (2009) suggested such a system with a single-start screw and a set of q near-full-complement of grooved rollers in which the set of drive grooves in each roller are consecutively axially shifted relative to one another by p/q for them to assume the same axial position relative to the nut. Since a nut cannot engage such staggered grooves, the rollers are provided at both ends with a small number of bearing grooves (on a smaller diameter) which are all in phase and supported in mating grooves in the casing; there being no *nut* as such. (There are also types with three-point support for higher rigidity in the radial direction; for the rollers not to bow). Roller spacing is maintained by their pintles inserted in two free guide rings. The rollers are referred to as stepped ones, and the system functions like a compound planetary gear set; a ring-less set with the screw and a sunless set with the casing, as shown in Fig. 10.7. Therefore, for one slip-free screw revolution, the guide ring (in lieu of a planet carrier) will rotate by $1/[1 + d_N d_{RS}/d_S d_{RN}]$ of a turn, in the same sense. The effective lead of the roller screw will be the one's complement of this value times the screw pitch. This, together with having a single-start screw thread, provides small values of the effective lead. The drawbacks of this system are its sensitivity to slip (just as with the threaded-roller screws) as well as the *uncertainty* of the effective lead, since it is function of the uncertain values of the mean-effective diameters of the mutually rolling surfaces.

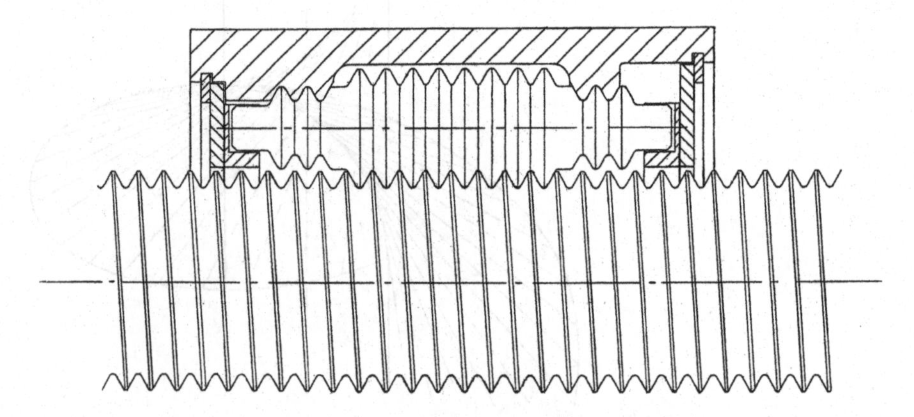

Fig. 10.7 Stepped grooved-roller screw with a single-start screw thread, showing one planetary roller of the set. The shown proportions give an effective lead of about $0.73p$

10.4 Rzeppa Joints of Higher Torque Density

A Rzeppa joint is a pivoting, rolling-element homokinetic joint capable of large operating angles between the shafts and of transmitting high torques. It has been invented by Rzeppa (1936) in his latest patent, which discloses an ingenious kinematic principle that keeps the set of balls to orbit in a plane bisecting the shaft angle; the condition for a constant-velocity joint. This design is essentially what is being used till now in a billion of units in automotive front wheel drives, to accommodate the steering as well as the vertical suspension movements. Rzeppa joints have customarily been designed with six balls, and are still being manufactured accordingly.

A suggestion was made by Sone et al. (2000), and later, of a Rzeppa joint with eight balls with an aim at increasing the torque density. For a given joint size, the balls are a little smaller in diameter and are arranged around the same pitch circle; the outside diameter of the cup member becoming somewhat smaller. Figure 10.8 shows a Rzeppa joint with eight balls in the position of maximum operating angle of 48°, divided in halves about the (vertical) median plane that contains the set of balls, in order for the side view to depict the actual orbit of the balls. The drawing further shows the equally offset centers of the ring-torus ball races in the inner and outer members, on their respective axes to opposite sides of the center, as well as the center of the ball orbit that will increasingly offset with increasing operating angle. The concentric double-spherical, slotted cage centers the two joint members, and allows for relative circumferential movement of the balls due to their eccentric orbit, but keeps them all in one plane. The latter task is a most essential contribution of the cage to a prompt execution of the angle bisecting action, rather than *expecting* each individual ball to follow the kinematics, all by itself.

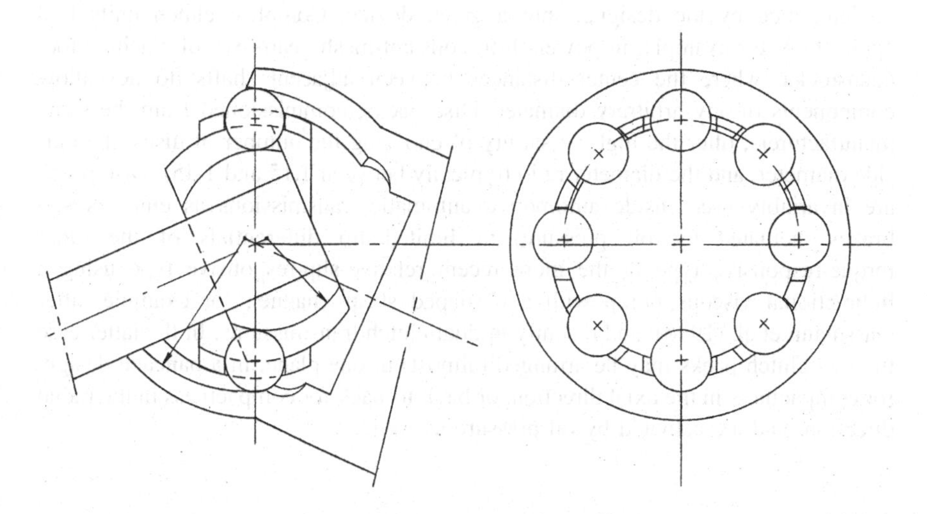

Fig. 10.8 Rzeppa joint with eight balls, shown in the position of maximum operating angle

10.5 Multi-Disc Clutches

Single-disc clutches with annular-shaped friction lining material on both faces of the disc have the merits of double friction surfaces, viz. double the torque density, requiring one axial actuating force that is applied to both surfaces, and the satisfaction of the design principle of nearest counteraction; the clutch disc being pressed between an axially fixed pressure plate and an actuated pressure plate in one enclosure. These merits have very early been recognized by car manufacturers; opting for the single-disc clutch to be placed between the engine and transmission (but it took them a half-century to start thinking of rendering their brakes the same virtues; replacing the drum brakes by disc brakes). A further advantage is that, when the clutch (like the brake) is actuated by fluid pressure, then the system will be self-compensating for lining thickness reduction due to wear.

The torque capacity of friction clutches can be increased, within a given diameter, by multi-disc design; clutches with packs of two to ten friction discs being commonplace. (These will have four to twenty friction surfaces). The pack alternately contains friction discs and plain steel discs; inner discs being connected to one terminal by a clearance-fit involute spline and outer discs being similarly connected to the other terminal, while the friction discs could be either of them. The two outermost steel discs could be disposed with when the pressure plates are assigned their function, so that a pack of ten friction discs is composed of 19 or 21 discs in total. Upon de-actuation the discs will back off each other, a process that could be enhanced by making the steel discs a little wavy. Increasing the torque capacity by that much and possibly reducing the outside diameter to operate at higher rotational speeds makes the multi-disc clutch of high-torque density.

Multi-disc clutches are very old and well-established machine elements that are provided by specialized manufacturers, either as complete units or as disc packs to be integrated by the designer into a given device. Complete clutch units find application, for example, in power-shift, constant-mesh gearboxes of machine tool headstocks, where the center distances between adjacent shafts do not allow components of any arbitrary diameter. Disc packs, commissioned from the same manufacturers, offer the high versatility of choosing the number of discs, the outside diameter, and the diameter ratio (typically between 1.15 and 1.45). Disc packs are invariably used inside automotive automatic transmissions as clutches and brakes, actuated by oil pressure, in limited-slip differentials of the older torque-responsive type or the more recent relative-slip-responsive type using a bidirectional viscous pump with a C-shaped shear channel, for example, after Gassmann et al. (1996), and recently in dual-clutch transmissions. In the latter case the two clutch packs may be arranged (almost) in one plane, in a pancake design, for compactness in the axial direction, or back-to-back for compactness in the radial direction, and are actuated by oil pressure as well.

10.6 Gear Pumps of High Performance Density

The pinnacle of gear pump technology is the design version with two helical *rotors*, each typically with seven rounded lobes, rather than with conventional involute gears. This design has originally been suggested by Morselli (2001), and was later adopted by a couple of world renowned hydraulics manufacturers. The typical appearance and proportions of the intermeshing rotors is shown in Fig. 10.9. The geometric problem to solve was to precisely specify a symmetrical-lobe profile in the transverse plane for two identical rotors, which satisfies the following objectives.

1. Keeping the two profiles in continuous single-point contact, including crest and root, as the rotors rotate oppositely at equal speeds. The contact point will move smoothly over the lobes along a stationary figure eight.

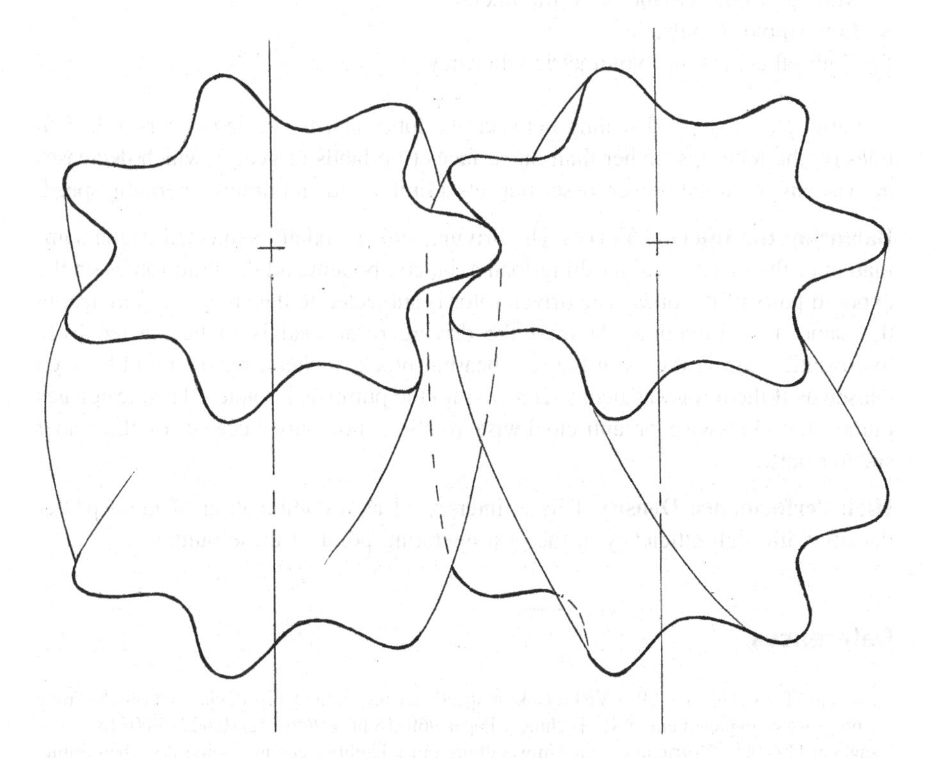

Fig. 10.9 Typical appearance of two intermeshing helical rotors of seven lobes, of a face advance of unity and an outside helix angle of 28°

2. Providing the largest possible fluid cell volume (between the lobes) for a given rotor center distance, in order to give the highest possible fluid volume pumped in one revolution (the displacement volume), hence power density of the pump.

The pressure angle between the two rotors in one transverse plane will vary between 90° (at the roots and crests) and zero somewhere about quarter heights, assuming medium values in-between. In order to have contact points of sufficiently small pressure angles at all times the lobes have to be helicoidal with a face advance of unity, so that the driving and driven surfaces will always be in line contact. These meshing characteristics render the pump superior characteristics regarding the following.

1. Almost completely eliminating the trapping of hydraulic fluid, hence the pressure spikes that would otherwise arise due to the compression of trapped fluid between the lobes; with negligibly small angular backlash.
2. Low noise emission.
3. Mitigated fluid cavitation at the inlet.
4. Low ripple or pulsation.
5. High efficiency and volumetric efficiency.

Labyrinth or stepped sealing between the lobes and the casing occurs at helical lines on the lobe tips, rather than on surfaces (top lands of gears), which decreases the viscous frictional power losses but sets a limit on the minimum operating speed.

Balancing the Internal Forces The driving rotor is axially subjected to the summation of the mechanical meshing-load axial component and the fluid forces on the exposed parts of its lobes. The driven rotor is subjected to the smaller difference, in the same axial direction. At least the driving rotor load is to be reacted by a hydrostatic bearing recess in the rear bearing block; its helix hand should be such chosen as if the rotor was being screwed into the pump as it rotates. This designates pumps for clockwise or anti-clockwise rotation, not only because of the wider suction port.

High-Performance Density This is interpreted as a combination of good power density with high efficiency at the best operating point of these pumps.

References

Gassmann T, Barlage J (1996) Visko-Lok: a speed-sensing limited-slip device with high-torque progressive engagement. SAE Technical Paper 960718.https://doi.org/10.4271/960718

Langbein U (2009) Vorrichtung zur Umwandlung einer Drehbewegung in eine Axialbewegung. DE Patent 10 2008 008 013, 1 October 2009

Morselli MA (2001) A positive-displacement rotary pump with helical rotors. EP Patent 1 132 618, 12 September 2001

Quaife RT (1985) Differential mechanism. EP Patent 0 130 806, 9 January 1985

Rzeppa AH (1936) Universal joint. US Patent 2,046,584, 7 July 1936

Saari OE (1966) Spin limiting differential. US Patent 3,292,456, 20 December 1966
Saari O (1986) Anti-friction nut/screw drive. US Patent 4,576,057, 18 March 1986
Sone K, Hozumi K, Kaneko Y et al (2000) Constant velocity joint. US Patent 6,120,382, 19 September 2000
Strandgren CB (1954) Screw-threaded mechanism. US Patent 2,683,379, 13 July 1954

Hydraulic Power Density

11

Fluid power transmission systems are known to allow the routing and conditioning of mechanical shaft power at best, fulfilling the largest variety of functional requirements in this respect. The inherent high power density of axial-piston-type hydraulic pumps and motors is compared for the swashplate and the bent-axis versions. Variable-displacement pumps and motors are required for the transmission ratio control—one of the most important functional requirements in power transmission systems—without reverting to fluid-throttling valve controls that "burn" power. The design of these variable-displacement units is given due consideration, especially regarding the configurations that promise higher efficiency.

11.1 Routing and Conditioning Mechanical Power

Mechanical power transmission systems are required to match an input mechanical shaft power to mechanical power consuming/dissipating systems. This matching viz. the routing and conditioning of mechanical shaft power is done with a view on one or more of several functional requirements, such as.

1. Direct connection, with or without misalignment accommodation.
2. Direction change, through a fixed or variable operating angle between the shafts.
3. Transforming; speed stepping up or down in a fixed ± ratio.
4. Reforming the output mechanical power into force × rectilinear velocity, at high effort or high speed.
5. Power splitting and/or recombining; rigid or differential.
6. Speed shifting; in discrete steps.
7. Speed control; continuously.

H. A. Arafa, *Design for Durability and Performance Density*,
https://doi.org/10.1007/978-3-030-56816-0_11

These functional requirements could be realized by either mechanical, hydro-mechanical, or electromechanical means. The latter two imply energy conversion twice; from mechanical power to and back from fluid power or electric power, respectively, in at least a part of the system. A comparison should be based on having an input mechanical shaft power and an output mechanical power in its rotary or rectilinear-motion version. Should the mechanical or the hydraulic power supply or generating system be included in the comparison, being put as a burden on the system, then the electric power supply or generating system should, might as well, be included. Some unfair comparisons may be based on taking for granted that electric power is just available by plugging into the grid, anywhere. This is true for stationary applications, but certainly not for airborne, seaborne, automotive and mobile equipment. Therefore, comparison is the least ambiguous when the power supply is disregarded. For deciding upon the most appropriate technology to adopt in any particular case, a comparative assessment is due regarding the following main attributes

1. Power density.
2. Energy efficiency.
3. Controllability, response speed.
4. Kinetic energy retrieval (recovery) potential.
5. Spatial extension (reach) capability.

While purely mechanical means for routing and conditioning power to achieve the afore-mentioned functional requirements is known to give the highest efficiency of all with a medium-to-high power density, it cannot compete with the other two technologies regarding controllability and response speed, as well as in systems of substantial spatial extension.

Hydraulic drives and actuators; motors and cylinders, are still of higher power density than their electromechanical counterparts; motors and electric cylinders (see Sect. 10.3). This is due to the properly applied principles of design for durability and performance density, and the properly selected material pairings, making hydraulic pumps and motors as well as cylinders endure operation under appreciably high fluid pressures. Electromechanical (electromagnetic) devices are prone to limitations on the electric current density in the windings, even in superconductors, and on the magnetic field strength. However, the energy efficiency of electric generators and motors has become a little higher than that of hydraulic pumps and motors.

11.1.1 Controllability and Efficiency

The following points are crucial for deciding to use hydraulics for a particular application regarding controllability and energy efficiency.

1. The now primitive way of using fixed-displacement pumps and motors with pressure relief valves and proportional control valves or servovalves for drive and control yields the worst system efficiency of all; it is simply a *burning the power to control it* scheme, in addition to it requiring effective cooling to dump the heat generated.
2. Better efficiency is achieved with variable-displacement pumps with pressure compensation, or in more elaborate circuits of load sensing. With pressure compensation the pump only produces the needed flow, but always under the set pressure. A load sensing pump provides the pressure and flow required, with a small additional pressure drop to be wasted. In either case, the pump operation at rated speed and less than full displacement volume reduces the efficiency due to too much rubbing with little power output.
3. One recently introduced remedy for some of these problems is the *variable-speed servo pump*, which will provide only the pressure and flow rate required by the actuator, with a small additional pressure drop, but without too much rubbing. In standby mode the fixed-displacement pump will run as slowly as to maintain leakage and pressure. Variable pumps could also be used to avoid turning the motor at too low speeds. The system is referred to as electro-hydro-mechanical, for the triple energy conversion involved. It relies on recent advances in electronic control technology and variable frequency drives with synchronous motors. Note that, with fixed pumps, all the controls are in the electronic motor supply and the hydraulic fluid is almost used as a rigid machine member, rather than being continuously imparted fluid power.

11.1.2 High Power Density Pumps and Motors

All the kinematic principles that create a fluid pumping/motoring function have been explored by inventors and manufacturers ever since the use of fluid power was envisaged. For achieving high power density the device should be capable of operating under high pressure without being subjected to unduly large internal reactions on sliding interfaces, and of handling large flow rates without suffering large interface sliding velocities. And for achieving controllability the device should lend itself to being designed as a variable-displacement one. These three requirements are best met with hydraulic pumps and motors of the (single-acting) piston type. These units are broadly classified into axial-piston and radial-piston types. The first could be sub-classified into inline and bent-axis types, while the latter into eccentric-shaft (single stroke per revolution) and multi-lobe cam ring types (see Sect. 2.3.2). Industry is continually refining the design of hydraulic pumps and motors, particularly those of the variable-displacement piston types.

11.2 Axial-Piston Pumps/Motors

A hydraulic axial-piston pump or motor consists of a cylinder block (barrel) that contains an odd number of axial bores inside which pistons are made to reciprocate in a (quasi-) simple harmonic motion in a sequential order, such that they fill-in the cylinder volume during half the cycle and discharge the fluid during the other half cycle of operation. The bores are connected to the back surface of the cylinder block through openings of a smaller cross-sectional area. The back surface remains in conformal contact with a valve plate with two crescent-shaped slots that communicate with the cylinder openings during filling-in and discharge; to function as a hydraulic distributor or commutator as the cylinder block and the valve plate are in relative rotation. (In all but the wobble-plate motors the cylinder block is the rotating member). The piston stroke motion is obtained by the piston ends or their rod ends being constrained to move in a plane inclined to the transverse plane of the cylinder block. The method of that constraining and the method of conversion between shaft power ($T.\omega$) and piston/rod power ($\Sigma F.v$) define the type of the pump or motor.

For the purpose of identification, simplified schematic drawings of the six basic types of fixed-displacement, axial-piston pumps/motors are given in Fig. 11.1. These types are the ones that are or have been (for some time) commercially available. The drawings do not include the casing or details such as seals, retainers, lubrication holes, or the bearing types. The basic pump/motor types are arranged in pairs under subheadings of *inline units* and *bent-axis units*, in an order of increasing the tilt angle responsible for producing the piston stroke, denoted α or β according to the type. This angle—besides other details—is primarily responsible for the achievable power density of the unit (see Sect. 11.2.1). Values of the number of pistons (N) and the (nearly maximum possible) tilt angle that are typical to the respective configuration are indicated. The bent-axis units require some synchronization means of the drive flange and cylinder block rotations, which are also seen in the drawings.

Variable-displacement versions of axial-piston pumps and motors are designed with various means for obtaining a variable swashplate tilt angle (α) in inline units or a variable cylinder block tilt angle (β) in bent-axis units, except for the type with bevel-gear synchronization. Worthiness to be designed to operate in an over-center mode will be discussed in Chap. 13.

11.2.1 Increasing the Power Density of Swashplate Units

A survey of the designs of axial-piston swashplate type pumps and motors reveals that they are customarily designed with the following features and proportions, as depicted in the partial schematic drawing in Fig. 11.2.

Inline units

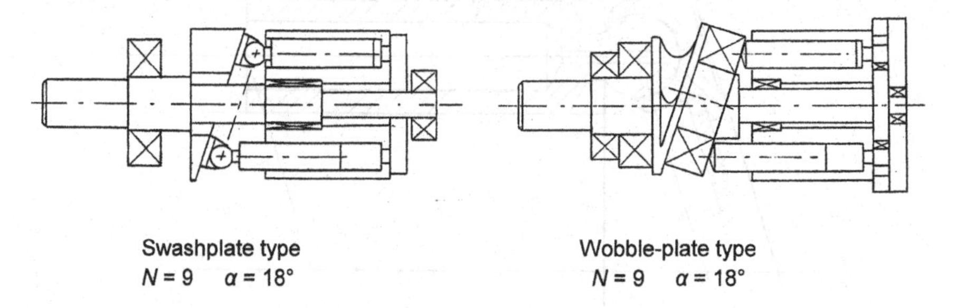

Bent-axis units of larger tilt angle

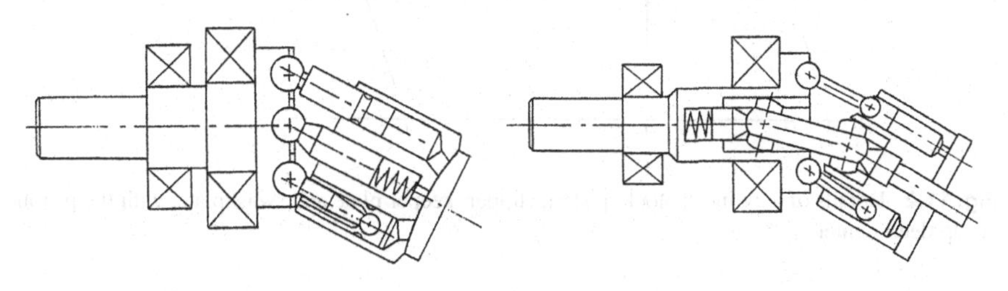

Bent-axis units of largest tilt angle

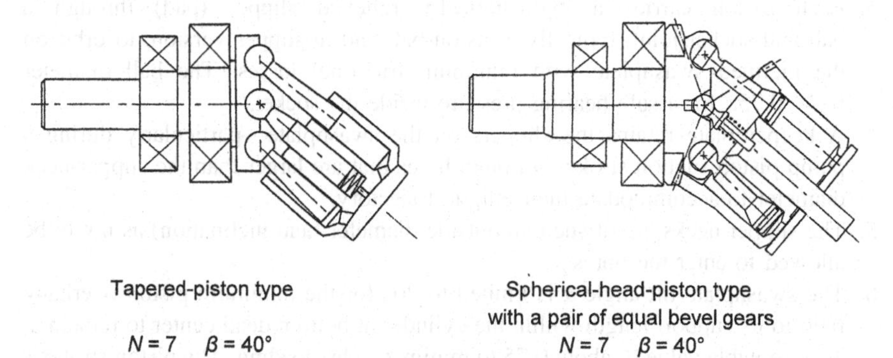

Fig. 11.1 Fixed-displacement, axial-piston pumps/motors

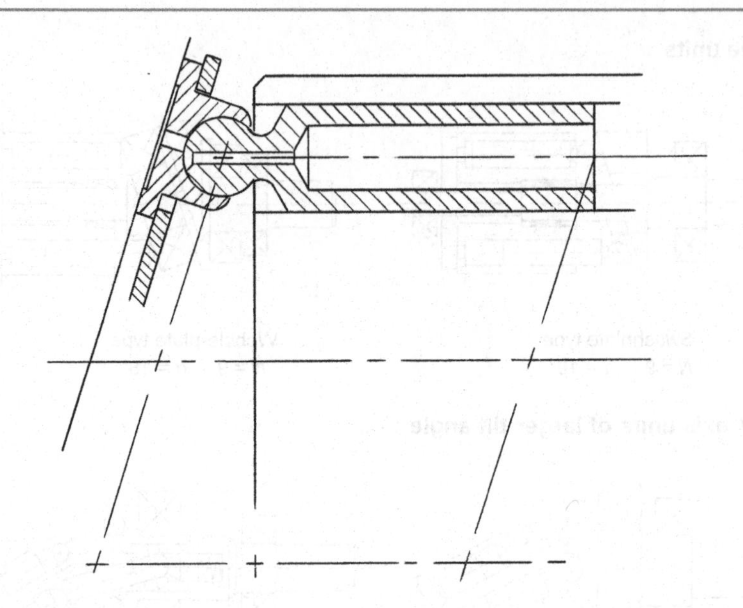

Fig. 11.2 Details of a cylinder block, piston, slipper, keeper plate and swashplate, with the piston at top dead center

1. Nine pistons, for high power density.
2. Pistons arranged on a pitch-circle diameter d_B in the cylinder block that is 3.75 times the piston diameter d to accommodate the array of slippers of sufficient base diameter (this ratio should be 4.55 for 11 pistons and 5.35 for 13 pistons).
3. Each piston carries a hydrostatically relieved slipper (pad) through a ball-and-socket joint, typically a piston-ball-end-in-slipper version, to orbit on the inclined swashplate with minimum frictional losses. The ball diameter is $\geq 0.75d$ for a safe bearing pressure inside the socket.
4. A keeper plate retains the slippers on the swashplate—particularly during a pump piston suction stroke—through holes that are larger than the slipper neck diameter to accommodate their elliptic trajectory.
5. The slipper necks are of such an outside diameter (and inclination) as not to be allowed to enter the bores.
6. The swashplate tilt angle α is limited to 20° for the minimum piston overhang ratio to its support length within the cylinder at bottom dead center to remain at an acceptable value of about 0.75 to minimize edge loading. The piston stroke is being given by $s = d_B \tan \alpha$.

A suggestion was made by Manring et al. (2013) to increase the power density of these units by inverting the piston-ball-end-in-slipper into a slipper-ball-head-in-piston design, which allows the ball center to approach the cylinder block face at piston top dead center, thus the angle α to be increased to 21°. Since this suggestion may necessitate a somewhat smaller ball diameter, the same goal could be achieved by reverting to the conventional design and letting the slipper necks intrude into the bores at piston top dead center. However, such suggestions are made in disregard of the design principle of *minimum pressure angle*, and they should further be scrutinized to ensure a proper trade-off between durability and power density.

11.2.2 Tapered-Piston Bent-Axis Units

A schematic of a tapered-piston, bent-axis pump or motor is shown in Fig. 11.3, without the centering pin that is required to maintain the intersection point of the axes of both rotating members. In this type the synchronization of the cylinder block with the drive flange is accomplished by the tapered pistons proper, which associate the functions of pumping and driving. The tapered piston shanks merge into spherical-segment heads that are in a close running fit inside the cylinders. These units are usually made with seven pistons and a cylinder block tilt angle $\beta = 40°$; as an industry standard. However, the drawing depicts a nine-piston unit for better interpretation of the tapered-feature synchronization; three pistons being momentarily in a driving orientation to the cylinder block, with a small pressure angle, in one direction of rotation, three others being in the opposite driving direction, and the remaining three assume too large pressure angles to be effective, as seen in the end view. Nine-piston units imply more slender pistons, as a comparison of the present drawing with the industry-standard designs would reveal.

Geometry The end view depicts equally spaced cylinder bores on a pitch-circle diameter d_B in the cylinder block, inside which are the projections of sections through the extremities of the tapered piston shanks, extended to their centers of articulation in the drive flange on a pitch-circle diameter d_F, the sections being taken perpendicular to the cylinder block axis. These sections are shown circular, but in fact they are very slightly elliptical, with the circle radius almost conforming to their radius of curvature at the oblong vertices in contact with the cylinder wall (virtual) extensions. Viewing the drive flange along the cylinder axial direction will show the pitch circle of the spherical joint centers of radius r_F projected as an ellipse. The major axis $a = r_F$ of this ellipse should be larger than r_B by the same amount the latter is larger than the minor axis of the ellipse $b = r_F \cos \beta$ to give equal amounts of outward and inward tilt of the piston axes from the cylinder bore axes, for maintaining straight-line contact of the piston shanks with the bores. Therefore.

$$d_F = 2d_B/(1 + \cos \beta)$$

The small tilt angle of the pistons in the cylinder bores reflects the ability to operate under high pressure without being subjected to large side thrust; an application of the design principle of minimum pressure angle.

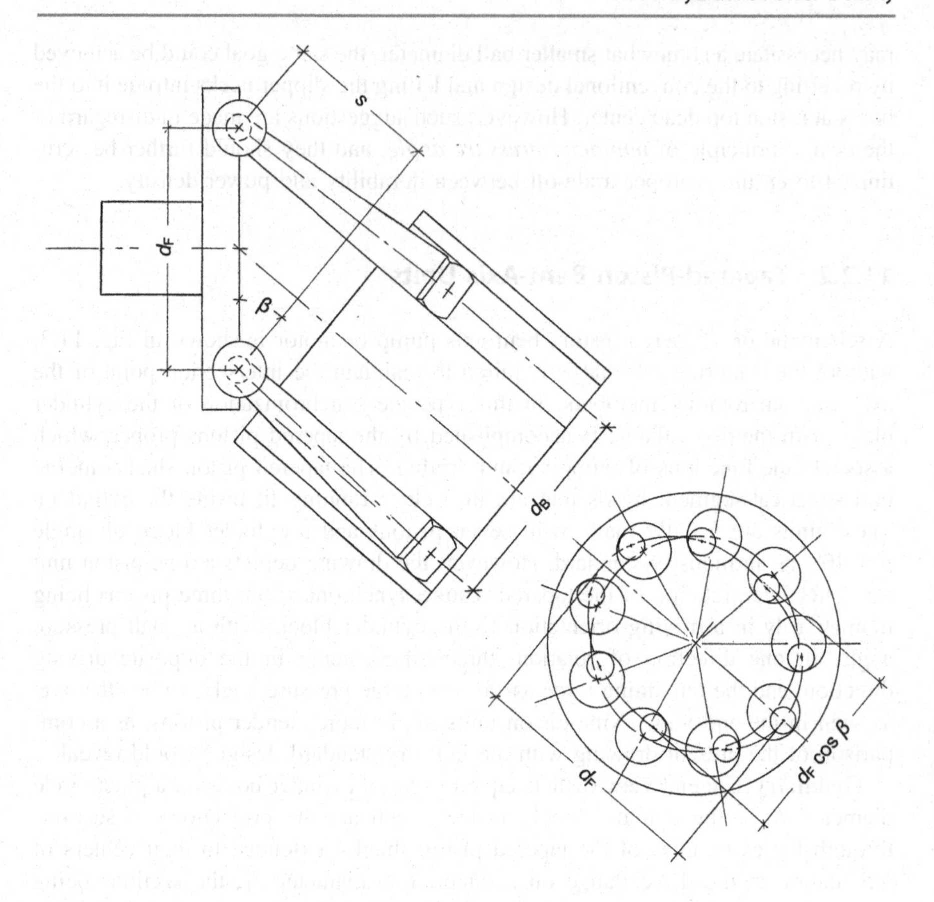

Fig. 11.3 Schematic and geometry of a tapered-piston, bent-axis pump/motor

Kinematic Geometry As the unit rotates, each tapered piston shank wobbles inside its cylinder bore, first assumed and later proven to be at the same rate but in an opposite direction. In Fig. 11.4 the (exaggerated) elliptic trajectory should be traced out by letting the eccentricity vector r_B of one bore center be at an angle θ at some instant, measured from the major axis, say, while letting the serially connected eccentricity vector e of one spherical joint center (from the bore center) be at the same contra-angle θ at the same instant. Therefore, the x- and y-coordinates of the end point of the two vectors will be.

$$x = (r_B + e)\cos\theta = a\cos\theta, \quad y = (r_B - e)\sin\theta = b\sin\theta$$

These are equations of an ellipse; the same rate is thus being proven. The drive flange is also proven to rotate at the same rate by recalling that the center of the one spherical joint just considered, existing at $x = a\cos\theta$, is in its actual position in the un-foreshortened direction in the projection.

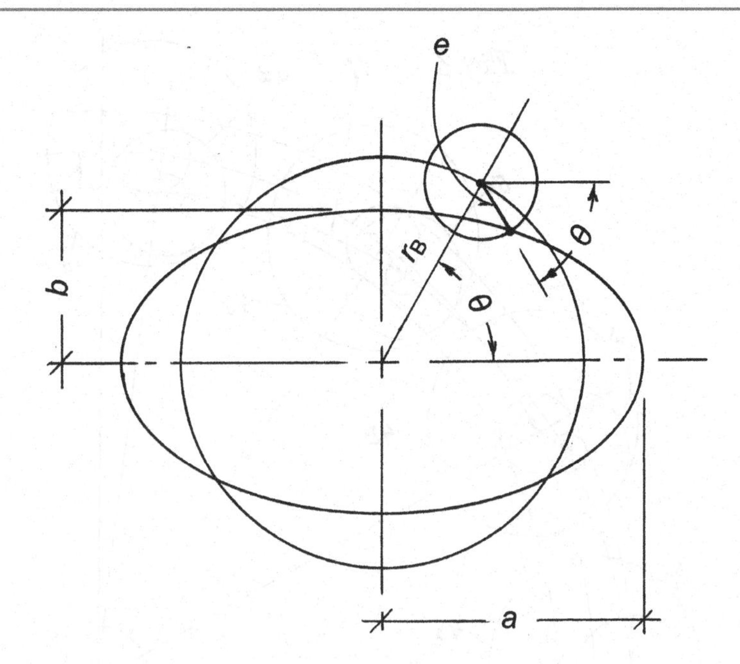

Fig. 11.4 Kinematic geometry of a tapered-piston bent-axis pump/motor

Tapered-piston bent-axis pumps or motors will not function according to the fine kinematic geometry just described unless the units are manufactured with absolute accuracy and operated in a perfect environment; unloaded and isothermal. This is for the assembly being kinematically highly overconstrained; it does not tolerate elastic deformations under load or differential thermal expansions of the parts, to the extent of jamming. This fact has already been recognized by Thoma et al. (1939), the inventors of the tapered-connecting-rod design; the forerunner of the present design after the same kinematic-geometric principle. (The original design has been manufactured into the 1980s, until the technique of making spherical piston rings was perfected, when it was replaced by the tapered-piston design). Then, and still now, remedy for the overconstrainedness was to make the shank of the piston (rod) slightly more tapered than theoretically evaluated for given geometric parameters. This makes the cylinder block lag behind the drive shaft, and the pistons (rods) to skew or windup; for a couple of them to bear on the inside cylindrical walls in a driving orientation (still maintaining straight-line contact), while others in or around the axial and lateral planes clear the walls. The original drawing of 1939 is reproduced in Fig. 11.5 to show the successive positions of the connecting rod smallest section (before the ball end) relative to the piston bore within about a quarter-sector of the cylinder block—as if the latter was exactly synchronized with the drive flange by separate means such that the clearances exist in radial direction—without revealing the lag, however (see also Sect. 13.2).

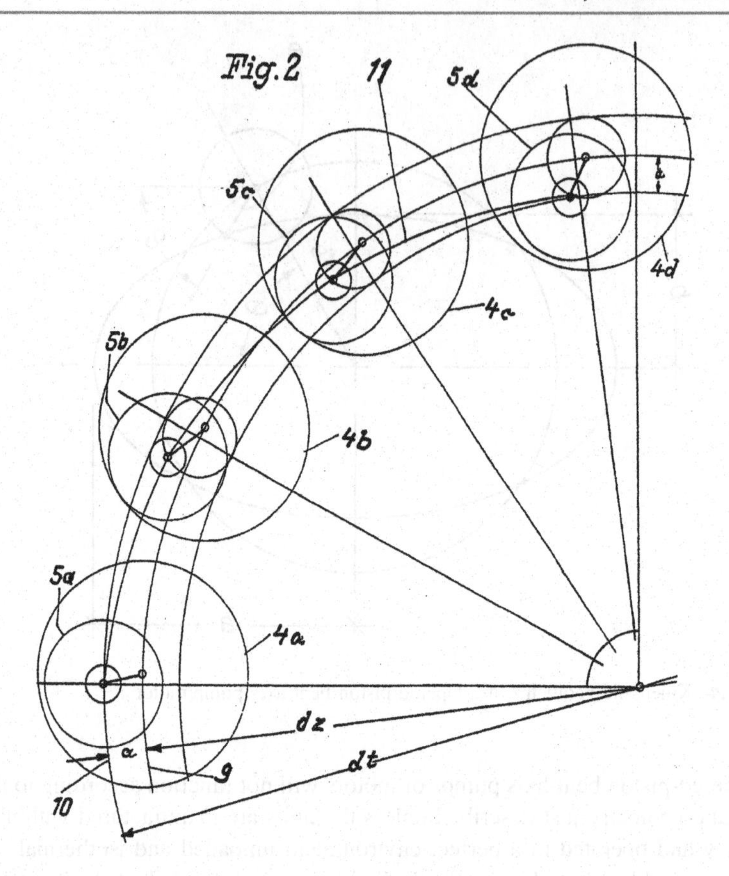

Fig. 11.5 Successive positions of the connecting rod smallest section relative to the piston bore. Reproduced from Thoma et al. (1939), public domain

Should the unit be a motor, to drive its load bidirectionally, then that lag will be present alternately to either side; revealed as lost motion between the cylinder block and the drive shaft. This feature would then (slightly) affect the exactness of timing of the valve plate, which is adjusted for one direction of rotation.

11.2.3 Comparing Same-Displacement Axial-Piston Units

The displacement volume (V_d) of a hydraulic machine is defined as the volume of fluid handled (under no pressure) in one revolution of the shaft. The stroke (s) hence the displacement volume (V_{dSW}) of a swashplate inline unit are expressed by.

$$s = d_B \tan \alpha$$
$$V_{dSW} = N(\pi/4)d^2 d_B \tan \alpha$$

where d is the diameter of the piston or the cylinder bore. For the typical number of pistons $N = 9$ the pitch-circle-to-piston-diameter ratio $d_B/d = 3.75$ to accommodate the array of slippers on the swashplate. For a swashplate tilt angle $\alpha = 18°$,

$$V_{dSW} = 9 \times (\pi/4) \times 3.75d^3 \tan 18° = 8.613d^3$$

The stroke (s) hence the displacement volume (V_{dBA}) of a bent-axis unit are expressed by.

$$s = d_F \sin \beta = (2d_B \sin \beta)/(1 + \cos \beta)$$

$$V_{dBA} = N(\pi/4)d^2 d_B (2 \sin \beta)/(1 + \cos \beta)$$

For the typical number of pistons $N = 7$ the pitch-circle-to-piston-diameter ratio $d_B/d = 3$ to leave sufficient material thickness between the bores. The typical tilt angle of the cylinder block $\beta = 40°$, so that.

$$V_{dBA} = 7 \times (\pi/4) \times 3d^3(2 \sin 40°)/(1 + \cos 40°) = 12.006d^3$$

This means that, for the same displacement volume, the swashplate unit should feature pistons of 1.117 the piston diameter of the bent-axis unit, arranged in the cylinder block at a pitch-circle diameter that is by about 1.4 times larger than in bent-axis units.

Typical constructions are shown to scale of two fixed-displacement axial-piston pumps or motors of equal displacement volumes; a swashplate unit in Fig. 11.6a and a tapered-piston, bent-axis unit in Fig. 11.6b. The drawings — to be considered only schematic — do not show external details of the casings or their back covers with port plates. But they do show essential features such as the biasing spring and its way of force application in the inline unit, as well as the abidance by the correct geometric proportions necessary for driving the cylinder block by the tapered pistons in the bent-axis unit.

Weight Comparison The relative sizes of the casings, cylinder blocks, and pistons evidence that the bent-axis unit is considerably lighter than the inline unit of the same displacement volume, mainly due to the larger angle.

Speed Comparison Operation as a motor is primarily considered in order to rule out the limitation due to cavitation in the pump suction route. Bent-axis motors could be operated at significantly higher speed than swashplate-type motors because of the much smaller sliding velocities at the interfaces hence the much smaller viscous frictional power losses, as well as the absence of axial inertia forces of the pistons as they orbit around a pseudo-cylindrical curtain.

Power Density Comparison The foregoing comparison items indicate a much higher power density of bent-axis units, especially motors, than is achievable with swashplate-type ones.

Cost Comparison Two factors decide the (much) higher cost of a bent-axis unit than an inline unit of the same displacement volume:

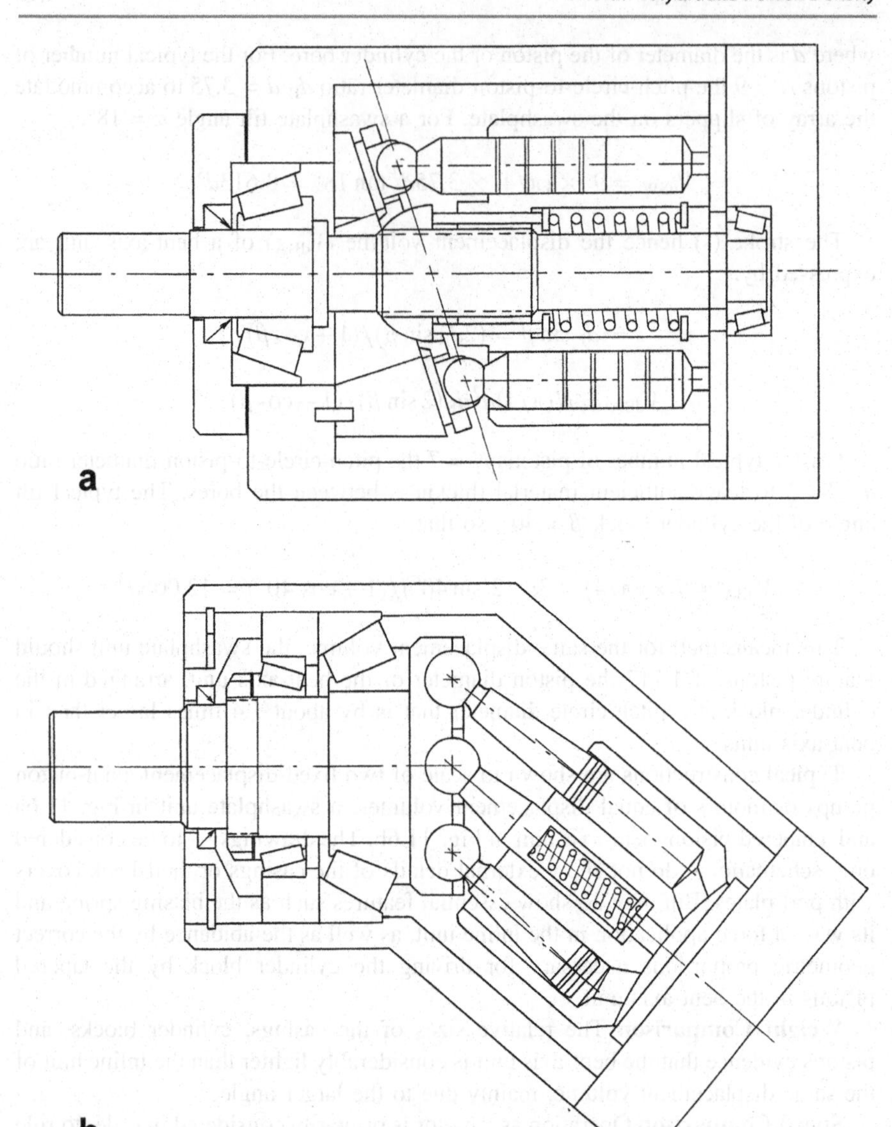

Fig. 11.6 Size comparison of axial-piston units of equal displacement volumes, **a** swashplate unit with $N = 9$ and $\alpha = 18°$, **b** tapered-piston, bent-axis unit with $N = 7$ and $\beta = 40°$

1. Mobility analysis reveals that bent-axis units of the tapered-piston type are kinematically highly overconstrained assemblies, which implies utmost dimensional and geometrical accuracy in their manufacture, with the associated cost penalty.
2. While the rolling bearings in swashplate units are not subjected to any appreciable loading (except from external shaft loading that could easily be avoided), the drive torque of bent-axis units is totally *taken* at the drive flange in form of piston forces that also produce heavy axial and radial loads, particularly on the first bearing. The size and quality of this bearing should be given particular attention when designing or commissioning a bent-axis unit, or else its life expectancy would be much at a disadvantage, and its slightest wear/looseness would collaterally damage the whole unit, because of the overconstrainedness.

11.2.4 Design Review of Pump Bearings

The Achilles' heel of bent-axis pumps and motors may be their rolling bearings; designers of these units have sometimes been mistaken about matching proper size bearings to the unit, to be commensurate with the data sheet announcement of the "maximum continuous operating pressure." The construction and the proportions of a fixed-displacement, bent-axis pump of the tapered-piston type that has been marketed for some time (in the past) was shown in Fig. 11.6b. The following design review calculations serve to reveal the bearing deficiency, if any. The pump design data are as follows.

Tilt angle of the cylinder block $\beta = 40°$

Number of pistons $N = 7$.

Piston diameter $d = 25$ mm.

Pitch-circle diameter in the cylinder block $d_B = 72$ mm.

Pitch-circle diameter in drive flange $d_F = 2\, d_B/(1 + \cos\beta) = 81.5$ mm.

Larger tapered roller bearing (A) designation: T7FC 070, with a large semi-cone angle of the cup, $\kappa \approx 30°$

Distance of support point from cup back face $a = 47$ mm.

Thrust factor $Y = 0.68$.

Basic dynamic load rating $C = 176$ kN.

Smaller tapered roller bearing (B) designation: 33,113.

Distance of support point from cup back face $a = 26$ mm.

Thrust factor $Y = 1.5$

Basic dynamic load rating $C = 142$ kN.

Bearings mounted in an O-arrangement; cups back-to-back.

Axial distance between the two cup back faces (maintained by a spacer sleeve) = 6 mm.

Distance between spherical-seat pitch plane and cone of bearing (A) back face = 21 mm.

Requirements

1. For a bearing life expectancy (with 90% reliability) L_{10} = 1,000 million revolutions determine the system pressure at which this pump could be operated.
2. Calculate the expected bearing life under a system pressure of 350 bar in continuous operation, just as the manufacturer's data sheets indicate.

Solution

Force Analysis on the Drive Shaft

The force acting on any piston on the pressure side is denoted F (= p_S × piston area). Force analysis is made with four of the pistons on the loaded side and the results are given in Fig. 11.7. These consist of the radial load F_r on the drive flange in the pump symmetry plane, the axial load on the drive shaft denoted F_e (the *externally* applied load on the bearing pair according to the terminology used in tapered roller bearing calculations), and the tilting moment on the drive flange due to the offset loads of the pistons on the loaded side.

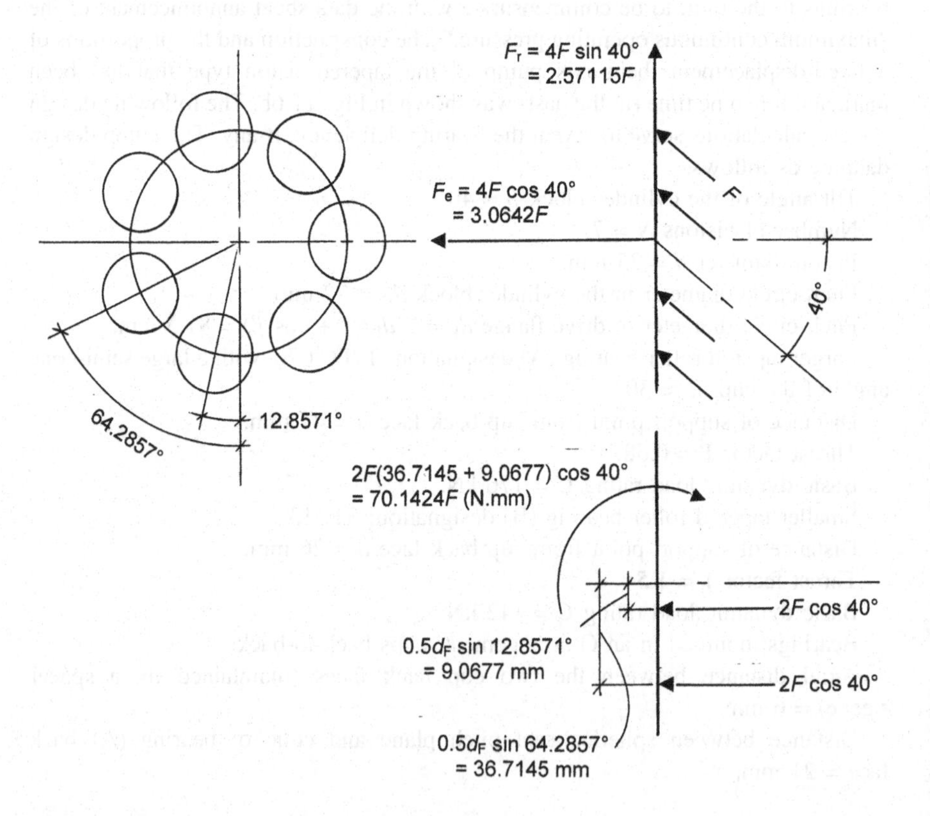

Fig. 11.7 Force analysis on the drive shaft

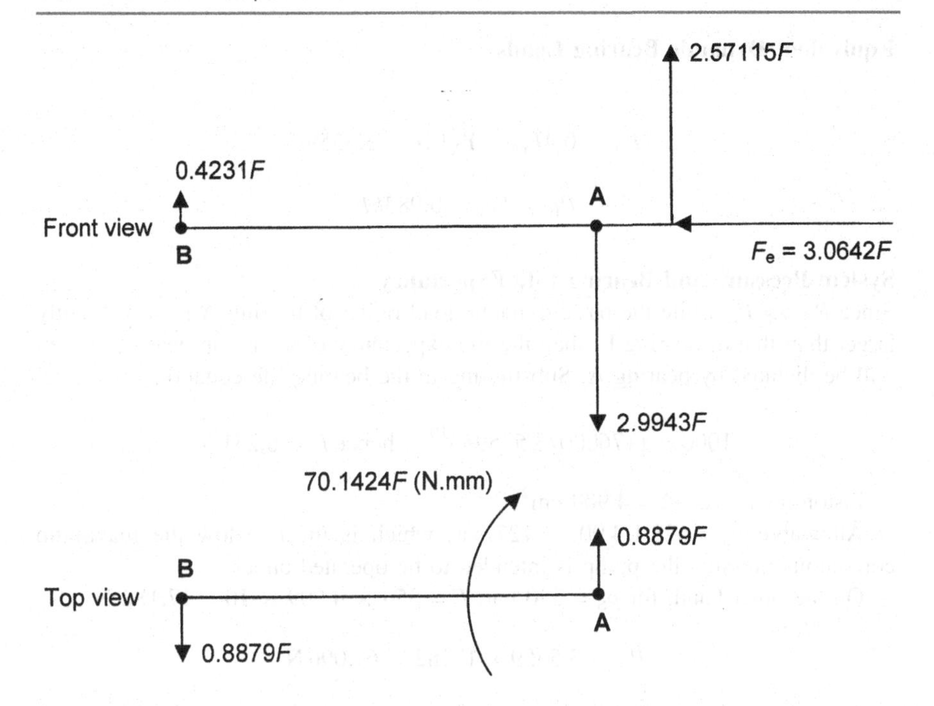

Fig. 11.8 Force analysis on the pair of tapered roller bearings in two orthographic views

Force Analysis on the Bearings

Bearing span = 47 + 6 + 26 = 79 mm, cantilever length = 39 + 21 − 47 = 13 mm, and the resulting load components on the two bearings are given in Fig. 11.8.

Resultant Bearing Loads

$$F_{rA} = \sqrt{(2.9943F)^2 + (0.8879F)^2} = 3.1232F$$

$$F_{rB} = \sqrt{(0.4231F)^2 + (0.8879F)^2} = 0.9836F$$

$$F_{rA}/2Y_A = 2.2965$$

$$F_{rB}/2Y_B = 0.3279$$

Loading condition: $F_{rA}/2Y_A > F_{rB}/2Y_B$; difference $\approx 2.14F$, while $F_e \approx 3F$.

Therefore, bearing B is partially circumferentially loaded and $F_{aB} = F_{rB}/2Y_B = 0.3279F$, while bearing A is fully circumferentially loaded so that $F_{aA} = F_{aB} + F_e = 3.3921F$.

Equivalent Dynamic Bearing Loads

$$P_A = 0.4F_{rA} + Y_A F_{aA} = 3.5559F$$

$$P_B = F_{rB} = 0.9836F$$

System Pressure and Bearing Life Expectancy
Since $P_A >> P_B$ while the basic dynamic load rating of bearing A is only slightly larger than that of bearing B, then the life expectancy of the pump bearing system will be dictated by bearing A. Substituting in the bearing life equation,

$$1000 = (176000/3.5559F)^{10/3}, \text{ hence } F = 6{,}231 \text{ N}$$

Piston area = $\pi d^2/4 = 4.909$ cm^2.

Allowable $p_S = 623.1/4.909 = 127$ bar, which is much below the maximum continuous pressure the pump is intended to be operated under.

On the other hand, for $p_S = 350$ bar, $F = 350 \times 4.909 \times 10 = 17{,}182$ N.

$$P_A = 3.5559 \times 17182 = 61{,}096 \text{ N}$$

$L_{10A} = (176{,}000/61{,}096)^{10/3} = 34$ million revolutions only, which could be reached in a few weeks.

Conclusion
The bearing selection for the pump at hand is far from being adequate. More recent pump or motor designs feature appreciably larger bearing proportions.

11.3 Performance of Hydraulic Pumps/Motors

The performance characteristics of any motor are usually plotted over a background of a torque–speed diagram that could include a set of so-called rectangular hyperbolic curves of constant power; $T \times \omega$ = constant. These curves are independent of the actual performance of the machine. A typical performance diagram (operating diagram) of a hydraulic motor is shown schematically in Fig. 11.9. In addition to the constant power curves it includes the following sets of main contour lines.

1. Substantially horizontal lines of constant values of the system pressure p_S.
2. Nearly vertical lines of constant values of the flow rate Q of the hydraulic fluid.

The first set of lines qualitatively shows a rather flat maximum torque at moderate speeds; little less torque being obtained at low speeds due to worse lubricating conditions and at high speeds due to viscous frictional losses. The second set of

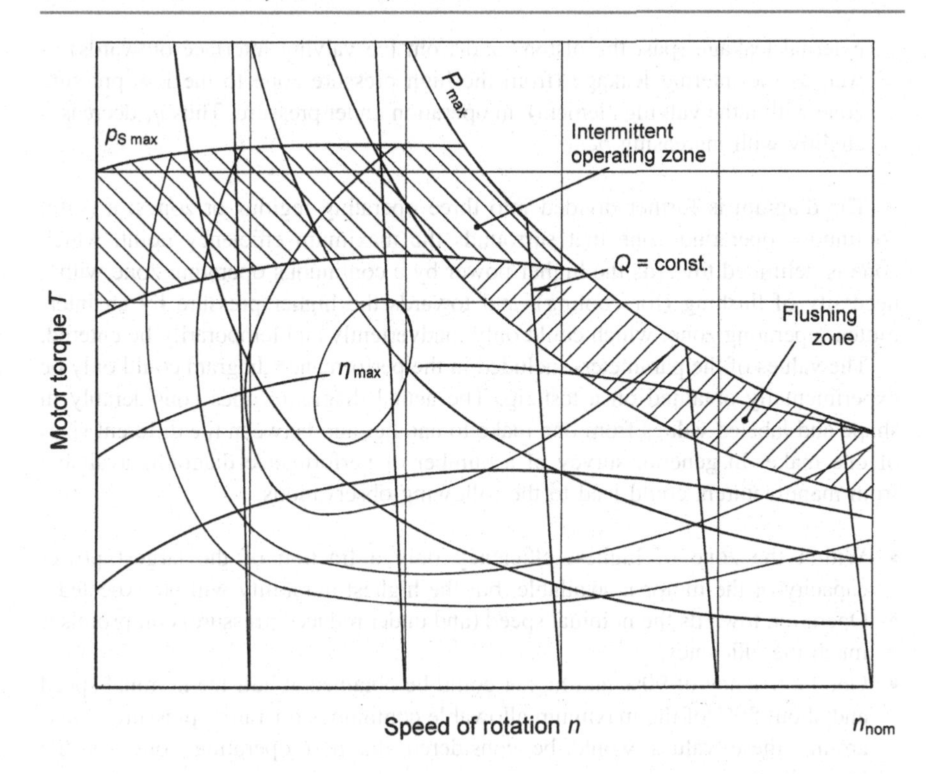

Fig. 11.9 Typical performance diagram of a hydraulic motor, unlabeled

lines is inclined "backwards" to indicate a slight decrease in speed as the system pressure increases, due to increased leakage. (In a performance diagram of a *pump* these lines will be oppositely curved and oppositely inclined). The abscissa is labeled up to the motor nominal speed n_{nom} which is the maximum allowable speed in continuous operation. (For some hydraulic motors a little higher maximum speed n_{max} is additionally specified for intermittent operation for a few seconds under reduced pressure). The diagram also includes the following.

1. A set of contour lines of constant (energy) efficiency η which surround an operating zone of highest efficiency η_{max}. (This efficiency is sometimes referred to as the total or the overall efficiency). The efficiency decreases as the operating point is chosen or moves away from this zone.
2. Not shown: a set of nearly horizontal lines of constant volumetric efficiency η_v of values slightly less than unity. The motor volumetric efficiency is the flow rate given by its displacement volume times its rotating speed divided by the (larger) actual fluid flow rate it takes in. This difference is mainly due to the

external leakage (past the pistons and from the valving interface outwards) as well as the internal leakage (from the high pressure zone to the low pressure zone within the valving element), in operation under pressure. Thus η_v decreases slightly with increasing p_S.

The diagram is further divided into three operating regions or zones: a (safe) continuous operating zone that surrounds the maximum efficiency point, which zone is delimited towards the higher power by a continuous operating zone with a necessity of flushing (for cooling), and towards the higher pressure by an intermittent operating zone, which could only inadvertently and temporarily be entered.

The values of the parameters included in the performance diagram could only be experimentally obtained on a test rig. The actual diagram varies considerably in shape and labeled values from one make to another and between the different sizes of one make. In general, survey of a number of performance diagrams available from manufacturers could lead to the following observations.

- Within the zone of highest efficiency only a fraction of the largest power capacity of the motor is available, but the highest durability will be expected.
- Operation towards the nominal speed (and under reduced pressure) compromises much the efficiency.
- Good efficiency of 90% and higher could be obtained at half the nominal speed and about 80% of the maximum allowable continuous operating pressure. Areas around these values would be considered the best operating zones of the hydraulic motor.

11.3.1 Energy Conversion

The relationships between the fluid pressure, flow rate, hence power, and the pump and motor parameters and operating variables are derived — based on the definitions of the displacement volume, efficiency, and volumetric efficiency of the two units. For a given amount of fluid power output from a pump or else input to a motor,

$$\begin{aligned} p_S Q &= T_P \omega_P \eta_P \\ &= T_M \omega_M / \eta_M \end{aligned} \tag{11.1}$$

Should this fluid power be transmitted under a given flow rate, then

$$\begin{aligned} Q &= n_P V_{dP} \eta_{vP} \\ &= n_M V_{dM} / \eta_{vM} \end{aligned} \tag{11.2}$$

Dividing Eq. 11.1 by Eq. 11.2 the pressure–torque relationships are obtained,

$$\begin{aligned} p_S &= (2\pi/V_{dP})T_P(\eta_P/\eta_{vP}) \\ &= (2\pi/V_{dM})T_M(\eta_{vM}/\eta_M) \end{aligned} \quad (11.3)$$

The *nine individual expressions* represented by Eqs. 11.1–11.3 are essential for calculations pertaining to a pump alone, a motor alone, or between the two units in a closed circuit. In the fluid power industry it is customary to express V_d in cm^3 and to use this value to specify the unit size. For calculations involving the torque, the displacement volume should be converted to m^3.

11.4 Higher-Efficiency Variable Pumps/Motors

In variable-displacement bent-axis machines the displacement volume is adjusted between zero and maximum by variably tilting the cylinder block relative to the drive shaft. Conventionally, the tilting axis passes through the intersection point of the drive shaft and cylinder block axes, being perpendicular to both. The machine will thus be referred to as center-pivoted. The tilting effort in either direction will be rather small. But it has the disadvantage that the pistons stroke around the cylinder mid-length, resulting in that the dead volume of pressurized fluid trapped at piston top dead center varies from a minimum at maximum stroke setting to more than half the stroke volume at the zero-displacement setting. Oil, being slightly compressible, stores an amount of energy per unit volume, under pressure. With a large dead volume under high pressure a considerable amount of energy is lost as the cylinders pass over from the high-pressure port to the low-pressure port; the compressed oil will instantly throttle out, noise will be generated, and the volumetric and energy efficiencies will decrease. The center-pivoted design is inevitable in variable-displacement machines in which synchronization of the cylinder block with the drive shaft is effected by a double Cardan shaft, a double tripod shaft, or a Rzeppa joint, where the distance between the pivot point and the cylinder block should remain constant. This design is also conventionally adopted in machines with tapered connecting rods or tapered pistons.

11.4.1 Off-Center-Pivoted Designs

A design in which the dead volume of oil is maintained at or near its minimum value (at maximum cylinder block tilt-angle) throughout the tilting range would contribute much to solving the aforementioned problem, hence to improving the efficiency in the range of less-than-full displacement volume. In tapered-piston units, the pivot point of cylinder block tilt could be placed at the center of the ball

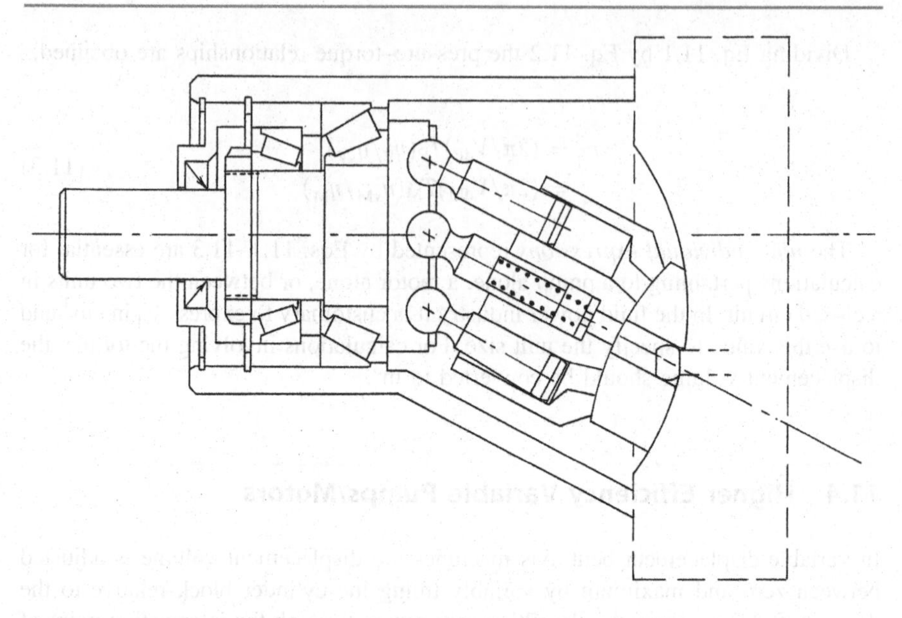

Fig. 11.10 Off-center-pivoted, bent-axis pump/motor with $\beta_{max} = 27°$

end of the pistons when in top dead center; an off-center-pivoted design, where the cylindrical inside surface of the casing should have its axis passing through the same point. Figure 11.10 depicts a seven-piston pump/motor according to that idea, where the sphero-cylindrical valve plate becomes asymmetrical as shown. However,

1. The spring-loaded centering pin moves outwards in its bore as the cylinder block tilts, providing a reduced holding-down force at maximum tilt. The maximum angle of tilt should therefore be rather limited, as shown. This problem could be mitigated in a nine-piston machine, where the pistons and the centering pin are longer hence a longer spring could be accommodated. (The limitation on the tilt angle is also set by the necessity that the two fluid slots in the valve plate back always cover the fluid ports in the casing).
2. Force analysis shows that the asymmetric valve plate has a decentering tendency; needing a larger de-tilting effort than in center-pivoted machines.

The problems of the limited range of cylinder block tilt angle and the inferior kinematics of synchronization of the cylinder block with the drive flange in variable-displacement units (see Sect. 13.3.1) could be resolved by some noteworthy design features suggested by Molly (1972, 1973). Figure 11.11 emulates those suggestions to highlight the basic ideas of an off-center-pivoted machine, which are (1) synchronization is achieved by a pair of identical bevel gears of toroidal pitch surfaces that roll over one another during tilting, allowing a maximum tilt angle of 45°, and (2) by necessity, the cylinder block is made to articulate about geared angle-bisector trunnion hinges (one on each side) in which the pinion pitch-circle radius equals the tube radius of the toroidal pitch surface of the bevel gears to keep their contact point; the momentary center of tilt, on the pitch point of the pinions.

The hinge links and the pinion hollow trunnions are configured as double rotary unions of ample-size passages to route the fluid between the tilting cylinder block and the fixed casing. The drawing further depicts a cradle that supports a flat valve plate, the cylinder block, and its spring-loaded centering pin. One of the hinge pinions is integral part of this cradle and the other is integral part of the fixed casing. This design does not allow the cylinder block to swing back to zero-displacement, a feature that is anyway not required in variable-displacement motors. However, the problem of large dead volume of fluid at piston top dead center still persists with the larger tilt angles. It should be remarked that toroidal bevel gears are not commonplace machine elements; they (still) lack viable machining process kinematics. But this design complies with the principle of minimum interface sliding per cycle (by virtue of the short pistons), the rotary group is in quasi-exact constraint (for high durability), and the large angle of tilt provides high power density without unduly inertially loading and skidding the rolling parts of otherwise needed constant-velocity joints, at high frequency and amplitude. For these features the present design could be expected to be re-considered at a later point in time, maybe by a manufacturer of fluid power transmission systems for wind turbines, as the generator-drive hydraulic motor.

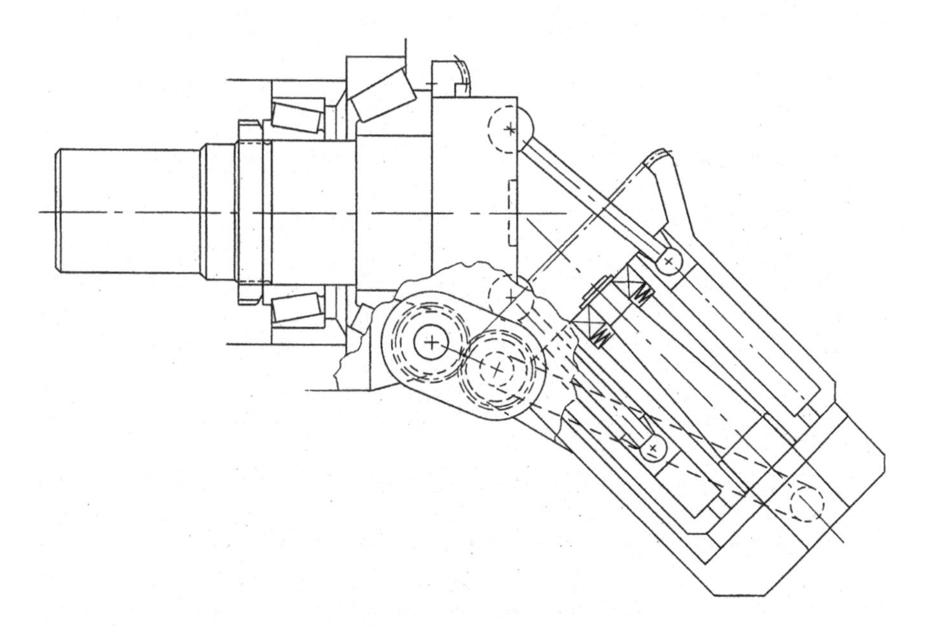

Fig. 11.11 Schematic of an off-center-pivoted unit with short pistons and synchronization by a pair of identical bevel gears of toroidal pitch surfaces, drawn without the casing

References

Manring ND, Mehta VS, Nelson BE et al (2013) Increasing the power density for axial-piston swash-plate type hydrostatic machines. ASME Transactions 135, J Mechanical Design, (July): 071002–1–071002–6. https://doi.org/10.1115/1.4023924

Molly H (1972) Die Axialkolben-Mehrzellenmaschinen in der Hydrostatik. VDI-Z 114(11):816–824

Molly H (1973) Axial piston type machine. US Patent 3,760,692, 25 September 1973

Thoma H, Molly H (1939) Hydraulic device. US Patent 2,177,613, 24 October 1939

Megawatt-Scale Fluid Power

12

Gearboxes are transformers of mechanical shaft power, and the larger ones could handle megawatt-scale power, without ratio controllability though. The latter attribute could be achieved by variable-displacement hydraulic pumps and/or motors. Multi-pump systems are described with an eye on wind turbine generator systems, with a detailed worked example to prove the feasibility and show the limitations on power of a variable-speed-ratio hydrostatic transmission system. Various wind turbine fluid power transmission layouts are presented and a comparative assessment thereof is made. The fluctuation or ripple in the fluid flow rate and torque in hydraulic pumps, and that in the shaft torque and angular velocity of motors of the piston type is discussed. The undesirable effect of ripple on the durability of the system components is highlighted, especially when dealing with high power. Mitigation of the ripple by angle phasing two units (pumps or motors), and units of back-to-back design is quantitatively assessed.

12.1 Transforming Mechanical Shaft Power

The transformation of megawatt-scale mechanical shaft power at a fixed speed ratio is not new to the mechanical engineering community. Examples of such transformers are large marine gearboxes for speed reduction and wind turbine gearboxes for speed increasing. The latter are used in wind turbine generator systems that include power electronics frequency converters to receive the electric power at variable frequency—due to the variation in wind speed—and output constant-frequency electricity to the power grid.

There are recent suggestions and endeavors to use fluid power transmission in wind turbine generator systems. One obvious advantage would be controllability, to eliminate power electronics with their inherent hazards and unreliability. The basic system would consist of a fixed-displacement pump of a large displacement volume, to be driven by the wind rotor and connected in a closed circuit to a

H. A. Arafa, *Design for Durability and Performance Density*,
https://doi.org/10.1007/978-3-030-56816-0_12

variable-displacement motor of a smaller displacement volume that directly drives a synchronous generator at its constant speed. The fluid power system thus performs a substantial part of the high-ratio speed stepping-up function, in addition to allowing controllability at best. Departing from the rated conditions, when the wind weakens, the pump discharges less flow rate under a smaller pressure, while the motor will be set by the control system to a smaller displacement volume to maintain the output speed yet under a smaller torque, hence smaller power. More elaborate circuitry and controls are still being developed around this basic system.

Still another advantage would be the possibility of placing the hydraulic motor, the electric generator, the fluid reservoir, and the cooling system at ground level; reducing much the volume and weight of the components in the nacelle (see Sect. 12.3.3).

12.2 Multi-Pump Systems

There may not be commercially available, hence well proven pumps of the size and to be operated at conditions such as described in Sect. 12.1. Therefore, suggestions were focused on using multiple pumps, to be driven from step-up gearing at speeds that bring them close to their best operating points. The rotor power is first input to a simple step-up stage with several output pinions, to *dump* the very high input torque by a twofold action: speed increasing and power splitting. The output torques of this first bull gear stage could then be handled by available more conventional means. And it is the design of this gear stage that should be given utmost attention, in particular regarding the means of accommodating elastic deformations and evading their effects that would otherwise lead to gear-tooth edge loading and breakage. However, the power capacity will be limited by that of the largest available variable-displacement fluid-power motor, which will be in the fractional megawatt range. Higher power could then be handled by arranging for two or more of such motors or systems in parallel. Recent suggestions of fluid-power transmissions for wind turbines according to these ideas were made by Basstein and Groenemans (2018) and De Vries (2016). The decision on opting for a large number of small pumps or else a smaller number of larger ones—may be of the same size as that of the motor—could be assisted by the results of the worked example in Sect. 12.2.1.

Multi-pump systems even out ripple in the discharge flow rate. But when one hydraulic motor is supplied with a ripple-free flow rate it will exhibit the inherent ripple characteristics in its output shaft speed and torque (see Sect. 12.4). With large multi-megawatt systems the consequences of the vibrations and high-frequency cyclic loading on the motor and generator may be grave; ripple should be given due attention at the design stage.

12.2.1 Worked Example

A hydrostatic transmission system for wind turbine generators of about 1 MW power rating is suggested as follows. A bull gear drives a number of pinions, each being followed by a planetary gearing for a step-up ratio sufficient to drive that number of hydraulic fixed-displacement pumps at their most appropriate speed, from the rated speed of the wind turbine rotor of 27 rpm. The discharge flow rate of the pumps is combined to drive one or more variable-displacement motor(s), which directly drives electric generator(s) at a synchronous speed of 750 rpm, for 50-Hz current.

Assumptions

One hydraulics manufacturer offers tapered-piston bent-axis pumps in the smaller size end in finely stepped sizes of 90, 106, 125 (cm^3), which have a nominal speed of 1,800 rpm, 1,700 rpm, and 1,600 rpm, respectively, and which weigh 23 kg, 27 kg, and 32 kg, respectively, and a largest pump of size 1000, nominal speed of 900 rpm, and weight of 330 kg. The largest available variable-displacement motor from the same manufacturer is of size 1000 and has a nominal speed of 1,600 rpm. The maximum continuous operating pressure of these units is $p_S = 350$ bar. For simplicity, assume also that the pumps and motors all have an efficiency of 0.925 and a volumetric efficiency of 0.98 at their operating points, and that the pumps should best be driven about 0.7 their nominal speed. Determine the following:

1. Motor best operating point; speed and system pressure for continuous operation, according to the information in Sect. 11.3.
2. Number of motors required.
3. Maximum required flow rate.
4. The operating point of the pumps to be chosen.

Trying firstly one of the small pumps, then the largest one, determine further the following:

1. Number of pumps to deliver the required flow rate (unrounded).
2. Actual number and drive speeds of the pumps, hence the speed step-up ratio.
3. Total weight of the pumps.
4. Input mechanical power to the pumps, assuming 100% gearing efficiency.

Solution

1. Motor operation at the synchronous speed of 750 rpm will be just under half the nominal speed of 1,600 rpm, thus under 0.8 the maximum continuous pressure for the best operating point. Therefore, $p_S = 0.8 \times 350 = 280$ bar

2. Maximum output power of one motor $P_{max} = T\omega$ $= [(p_S V_{dM}/2\pi)\eta/\eta_v](2\pi n/60)$ (according to Eq. 11.3) $= (280 \times 10^5 \times 1000 \times 10^{-6}) \times (0.925/0.98) \times (750/60) \times 10^{-3} = 330$ kW. Therefore, three motors will be required for an output power of 990 kW.
3. Maximum fluid flow rate $Q_{max} = n_M V_{dM}/\eta_v = 3 \times 750 \times 1/0.98 = 2{,}296$ L/min
4. The pumps will be operated under 280 bar at about 0.7 their nominal speed; $Q \approx 0.7 n_{nomP} V_{dP}\, \eta_v$

First, Try Pump Size 106

1. $Q = 0.7 \times 1700 \times 106 \times 10^{-3} \times 0.98 = 123.617$ L/min per pump
   Number of pumps = 2296/123.617 = 18.5735.
2. Actual number rounded down to 18.
   Drive speed = 0.7 × 1700 × 18.5735/18 = 1,227.9 rpm.
   Overall speed step-up ratio = 1227.9/27 = 45.478.
3. Total weight of pumps = 18 × 27 = 486 kg
4. Input mechanical power $P = (2\pi n/60)[(p_S V_{dP}/2\pi)\eta_v/\eta]$
   $= (1227.9/60) \times 280 \times 10^5 \times 106 \times 10^{-6} \times (0.98/0.925) \times 10^{-3}$
   = 64.35 kW/pump.
   Total power = 18 × 64.35 = 1,158.3 kW.

Second, Try Pump Size 1000

1. $Q = 0.7 \times 900 \times 1000 \times 10^{-3} \times 0.98 = 617.4$ L/min per pump

   Number of pumps = 2296/617.4 = 3.7188.
2. Actual number rounded up to 4.
   Drive speed = 0.7 × 900 × 3.7188/4 = 585.71 rpm.
   Overall speed step-up ratio = 585.714/27 = 21.693.
3. Total weight of pumps = 4 × 330 = 1,320 kg
4. Input mechanical power $P = (2\pi n/60)[(p_S V_{dP}/2\pi)\eta_v/\eta]$
   $= (585.71/60) \times 280 \times 10^5 \times 1000 \times 10^{-6} \times (0.98/0.925) \times 10^{-3} =$ 289.58 kW/pump.
   Total power = 4 × 289.58 = 1,158.3 kW.

Check: In both cases the system overall efficiency is 3 × 330/1158.3 = 0.855 (= 0.925^2).

Conclusion The choice of 18 small pumps makes the system of much higher power density than the choice of four large pumps. In either case the speed step-up ratio cannot be achieved by the bull gear stage alone; add-on gearing will be required.

12.3 Wind Turbine Fluid-Power Transmission Layouts

Closed-circuit fluid-power transmission systems of the megawatt-scale could be used in horizontal-axis wind turbine generator installations according to a variety of conceptual layouts. Starting with the mostly compact layout, two further ones are presented in the order of increasingly relocating heavy system components down from the nacelle to the ground level. This action should significantly facilitate and reduce the expenses of monitoring, maintaining, and replacing these components when necessary.

12.3.1 All the System Components in the Nacelle

The *substantial* components as shown in Fig. 12.1, are the pump(s) with their necessary step-up gearing, the motor(s), the hydraulic fluid reservoir, the charge pump that makes up for the external leakage of the pump and motor into the low-pressure line, the cooling system, and the electric generator(s). The only advantage of such a layout is shortest hydraulic and mechanical connections between the components, viz. compactness. The disadvantages are a heaviest-of-all nacelles and that the electrical power cable through the tower will be twisted when the nacelle yaws to continually face the wind, so that after a couple of turns the nacelle will have to be rotated back to its home position, in a production stop period.

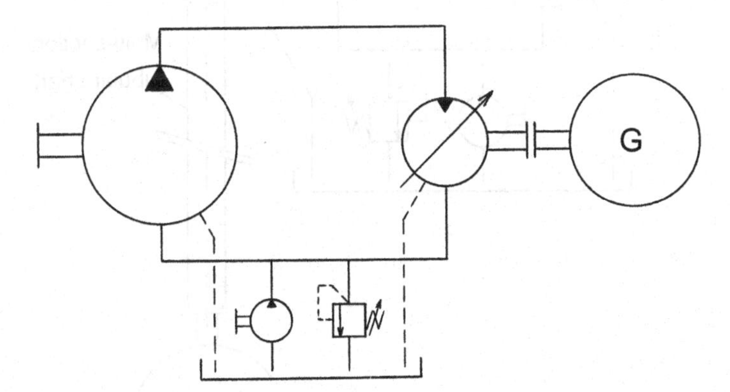

Fig. 12.1 Simplified layout of the most compact fluid power transmission for a wind turbine generator system

12.3.2 Hydraulics in the Nacelle, Generator at Ground Level

This layout as shown in Fig. 12.2 may not have been suggested before, but it might be worth considering. In addition to the pump(s) on a horizontal axis, the hydraulic motor would be mounted vertically in the nacelle to achieve a 90° direction change with the disposition of the fluid power units, without using bevel gears. The motor drives the electric generator that is also vertically mounted inside the tower base through a long shaft. The advantages would be shortest hydraulic connections in the nacelle and shortest electrical connections on the ground level. The drawback is that a long multi-section tubular shaft (such as described in Sect. 10.1) is required for power transmission, to be guided and supported inside the tower in several radial and thrust bearings along its way down to the generator. The thrust bearings should be supported on adjustable compliant elements to take equal shares of the weight; in a cascade. Another probable drawback would be a need for a right-angle gearbox for the generator not to be installed inside the tower base.

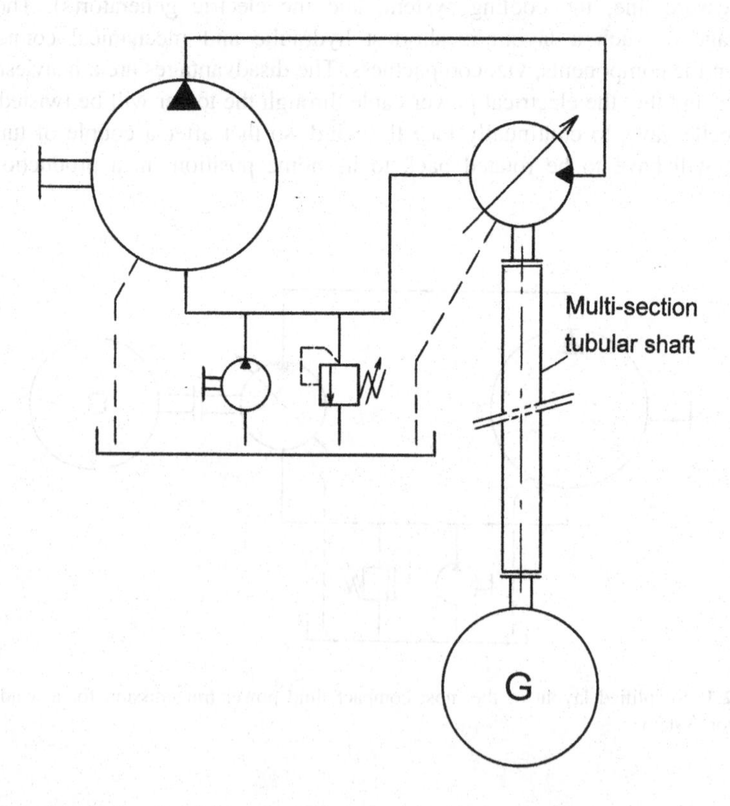

Fig. 12.2 Schematic of a fluid-power plus mechanical-shaft transmission system for a wind turbine generator system, with the generator brought down to ground level

12.3.3 Pump(s) Only in the Nacelle

Such a layout was suggested by Dahlhaug (2013), and is schematically shown in Fig. 12.3. It requires a swivel joint of multiple channels on top of the tower for the tubing with the ground-level components not to twist. In addition to the two channels of the closed circuit fluid power transmission, a third one needed for the pump case drain; to bring the pump external leakage oil down to the reservoir. (A fourth channel would be required if flushing the pump casing was indicated). The shown pancake configuration of the swivel joint should only be considered symbolic; such an application requires a cylindrical-interface design similar to the hydraulic distributor described in Sect. 5.1.4 and Fig. 5.5. The advantages of

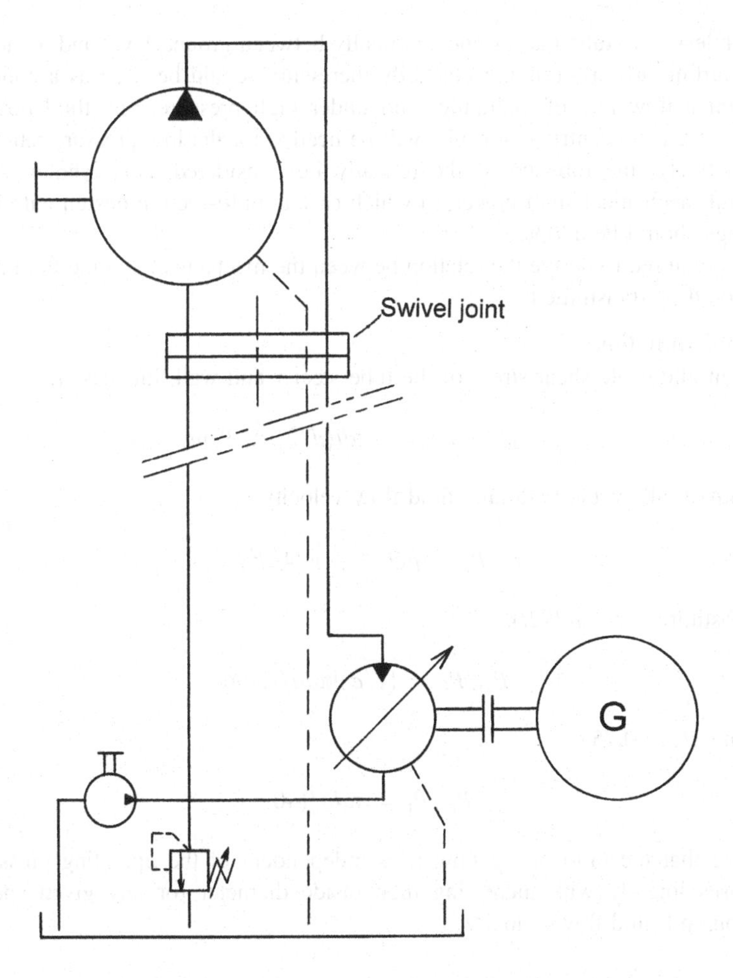

Fig. 12.3 Schematic of a fluid power transmission for a wind turbine generator system with the pump only in the nacelle

putting the pumps only in the nacelle are lightest of all nacelles and short mechanical and electrical connections. One disadvantage is that the charge pump should become a charge/boost pump since its pressure should be higher by about 15 bar to sustain the hydrostatic head of the tower height, in addition to some 5 bar to overcome the pipe-flow frictional loss. Another disadvantage is that the rotary union and the tubing itself should be of relatively large sizes, since a power of 3 MW, for example, under a system pressure of 300 bar requires a flow rate of the hydraulic fluid of 100 L/s, and this flow needs to be routed with a minimum possible pressure drop.

12.3.4 Comparative Assessment

A seamless steel tube that extends vertically between ground level and the top of a wind turbine tower—running coaxially therewith—could be used as a conduit to transmit a flow rate of hydraulic fluid under high pressure, viz. fluid power, in which case a concentric outer tube will be needed for the low pressure return line. Alternatively, the tube could theoretically be considered as a tubular shaft to transmit mechanical-shaft power, in which case a multi-section design with hanger bearings should be adopted.

It is required to derive the relation between the mechanical and the fluid powers that could be transmitted.

Power Estimation

With an allowable shear stress of the tube steel τ and wall thickness t,

$$P_{\mathrm{m}} = T\omega = \tau \pi d t (d/2)(\pi/30) n$$

With an allowable hydraulic fluid flow velocity v,

$$P_{\mathrm{f}} = pQ = p(\pi/4) d^2 v$$

Substituting $\sigma = pd/(2t)$,

$$P_{\mathrm{m}}/P_{\mathrm{f}} = (\tau/\sigma)[\pi d n/(30 v)]$$

With $\tau/\sigma = 0.75$,

$$P_{\mathrm{m}}/P_{\mathrm{f}} = \pi d n/(40 v)$$

Note that the ratio of the powers is independent of the operating pressure. It increases linearly with increasing tube inside diameter for any given speed of rotation and fluid flow velocity.

Numerical Example

n = 1,000 rpm, v = 4 m/s, p = 300 bar, τ = 300 MPa, d = 100 mm, hence t = 3.75 mm. Therefore, P_m = 1.8506 MW and P_f = 0.9425 MW.

With a larger d = 140 mm we get t = 5.25 mm, and the transmittable powers become $P_m = (1.4)^3 \times 1.8506 = 5.078$ MW, while $P_f = (1.4)^2 \times 0.9425 = 1.847$ MW, with the ratio between them ≈ 2.75.

Conclusion

Bringing down mechanical-shaft power would be a better solution than bringing down fluid power, especially for multi-MW power ratings.

12.4 Ripple in Piston Pumps and Motors

Hydraulic piston pumps deliver a flow rate which is the summation of the volumetric displacement rates of the pistons that simultaneously connect with the discharge half of the pump. When the pump is driven at a constant speed of rotation it will be expected that its discharge flow rate will ripple or fluctuate at some frequency that relates to the piston passing frequency, and that the driving torque should suffer the same. Assume that the piston motion is purely sinusoidal (such as in axial-piston swashplate pumps and in radial-piston pumps driven by an eccentric polygon ring), and that the discharge stroke fully occupies half a revolution. The resulting flow rate summation is shown in Fig. 12.4a for the odd-numbered pistons of five, and in Fig. 12.4b for the even-numbered pistons of six.

The sum discharge flow rate is obtained as a succession of sinusoidal caps. When each cap represents the contribution of three pistons (for $N \leq 8$), and the amplitude of the individual sine curves is set to unity, the equation of one cap as function of the rotation angle θ will be.

$$Q = \sin\theta + \sin[\theta + (360°/N)] + \sin[\theta - (360°/N] = [1 + 2\cos(360°/N)]\sin\theta$$

which is a sine curve, in phase with the middle (reference) sine curve, of the same half wave length (the dashed extension of the sine curve in Fig. 12.4), and of an amplitude that increases with increasing N. This amplitude is the maximum instantaneous flow rate of the pump (Q_{max}). For $8 < N \leq 12$ two further sine curves count, and so forth. The above equation of the sum discharge flow rate is valid in a range $\theta = 90° \times (1 \pm 1/N)$ for odd N, or $\theta = 90° \times (1 \pm 2/N)$ for even N.

By virtue of the number of pistons, being odd or even, the ripple is obtained at twice the piston passing frequency in the first case and at the piston passing frequency in the second. Taking a cap vertex (middle) as an origin, the minimum flow rate at the cusps will thus be given by

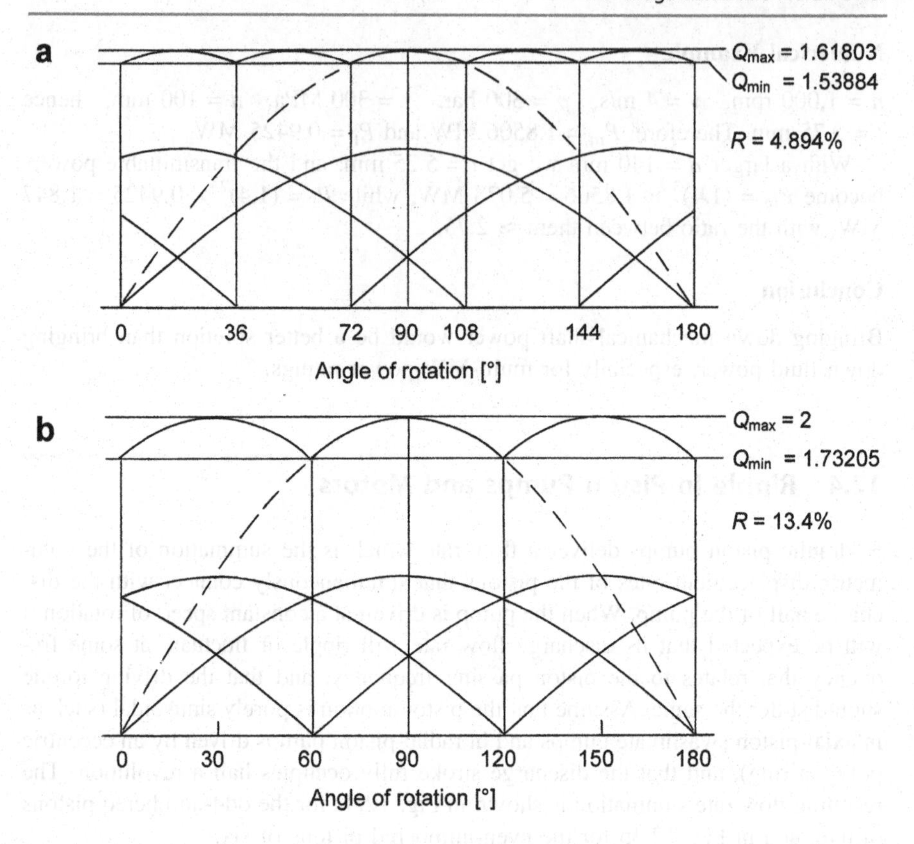

Fig. 12.4 Discharge flow rate and ripple, **a** for a five-piston pump, **b** for a six-piston pump

$Q_{min} = Q_{max} \cos(90°/N)$ for odd-numbered pistons
$Q_{min} = Q_{max} \cos(180°/N)$ for even-numbered pistons.

Defining the flow ripple R as the difference between Q_{max} and Q_{min} divided by Q_{max} then.

$$R_{odd} = 1 - \cos(90°/N) \tag{12.1}$$

$$R_{even} = 1 - \cos(180°/N) \tag{12.2}$$

Equations 12.1 and 12.2 have already been given by Lezard (1964); they must have been derived earlier. A graphical plot of their results in Fig. 12.5 shows that pumps with even-numbered pistons have an unacceptably high ripple compared with those having odd-numbered pistons.

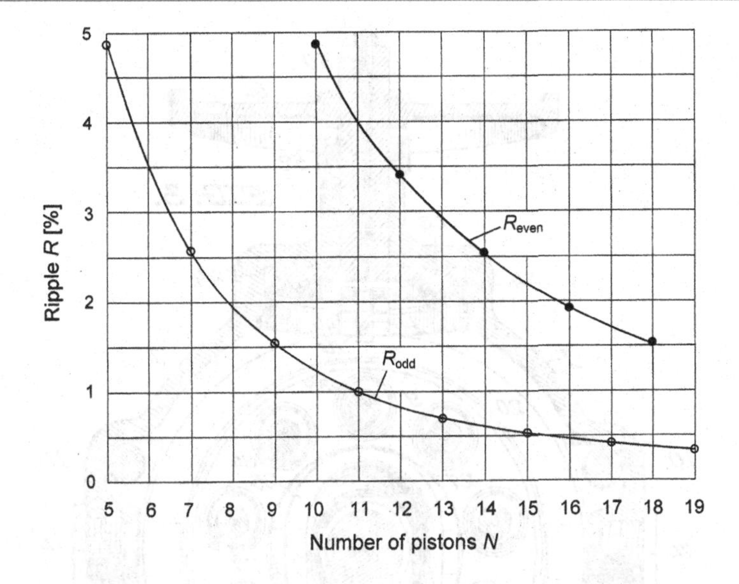

Fig. 12.5 Ripple of piston-type units of odd and even numbers of pistons

The oldest evidence that hydraulic piston-type machines should feature an odd number of pistons "for smooth operation" is furnished by Janney (1909), of which the original drawing is reproduced in Fig. 12.6. He also discovered, in the same work, that the ripple of a 9-piston machine is the same as that of an 18-piston one, in accordance with the above two expressions of the ripple.

Ripple Effects A pump operated at a constant speed against a constant system pressure will reflect its discharge flow ripple on its drive torque, in phase with it. A hydraulic motor supplied with a constant pressure and a constant fluid flow rate (constant power) should, theoretically, exhibit ripple in its shaft speed and its output torque, oppositely oriented or in anti-phase. In practice, however, the mass moment of inertia of the motor and its connected load would largely prevent speed ripple, leaving only the torque to ripple, and the motor intake flow rate will consequently ripple. In any case torque ripple is a source of cyclic loading or fluctuating stressing on the drive shaft and other power transmission components, which could lead to fatigue failure if not properly anticipated. This is particularly important in five-piston units. But the fluid compressibility reduces the effect of ripple on the system; the sharp cusps of the curves in Fig. 12.4 should not actually be so.

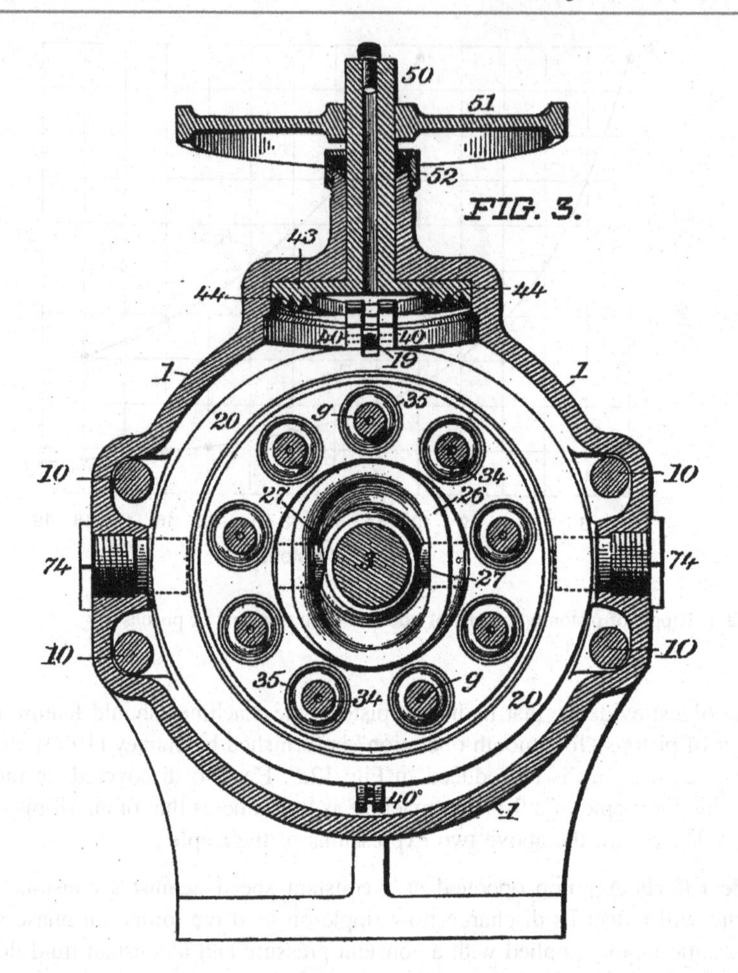

Fig. 12.6 Oldest evidence that hydraulic piston machines should be of an odd number of pistons. Reproduced from Janney (1909), public domain

12.4.1 Optimum Number of Pistons

In hydraulic axial-piston pumps and motors the cylinder bores are equally spaced in the cylinder block on a pitch circle of diameter d_B, Fig. 12.7a. The minimum material thickness between the bores and the block outside surface is typically one quarter the bore; 0.25*d*, to make the elastic deformation under pressure negligibly small. The minimum material thickness between the adjacent bores should rather be 0.28*d* in order to allow slipper pads of sufficient size in swashplate units and piston ball ends of sufficient diameter in bent-axis units. With any given tilt angle of the swashplate, or the cylinder block, the piston stroke *s* is taken as a constant fraction *k* of the pitch-circle diameter. For a given cylinder block diameter *D* it is required to

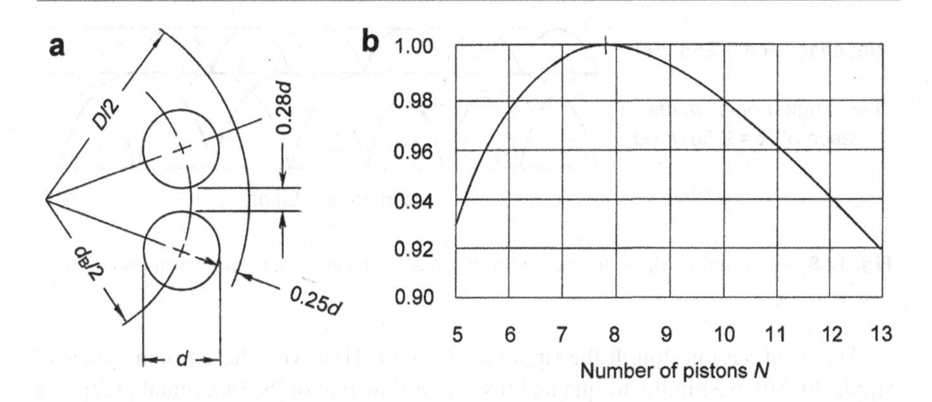

Fig. 12.7 Optimum number of pistons for maximum displacement volume, **a** proportions, **b** plotted function

determine the optimum number of pistons N that produces a maximum displacement volume V_d, hence to make the unit of highest power density.

$D = d_B + 1.5d$

$d_B \sin(180°/N) = 1.28d$

$s = kd_B$

$V_d = (\pi/4)d^2Ns$

$V_d = KD^3N[\sin^2(180°/N)]/[1 + 1.172\sin(180°/N)]^3$

where K is another constant of proportionality. As shown in Fig. 12.7b this expression in N has its maximum value (set to a reference of unity) at $N = 7.8$. The characteristics are rather flat, and small changes in the assumed minimum material thicknesses do not affect the result much. It is known that the majority of pumps and motors have seven or nine pistons; they should thus be of highest power density.

12.5 Mitigating Ripple by Angle Phasing Two Units

Hydraulic piston-type pumps and motors are invariably made with an odd number of pistons for the benefit of having a small ripple. The net effect of ripple could further be reduced by operating two identical units hydraulically in parallel, geared up at the same speed, when adjusted to a relative phase angle of 360°/4 N. The two ripple waves will be phase shifted by the same angle, and a total of twice the number of cusps/caps will be obtained as shown in Fig. 12.8. Superposition will give a net ripple wave of twice the frequency, i.e., four times the piston passing frequency, and the ripple will be obtained as.

$$R = 1 - \cos(90°/2N)$$

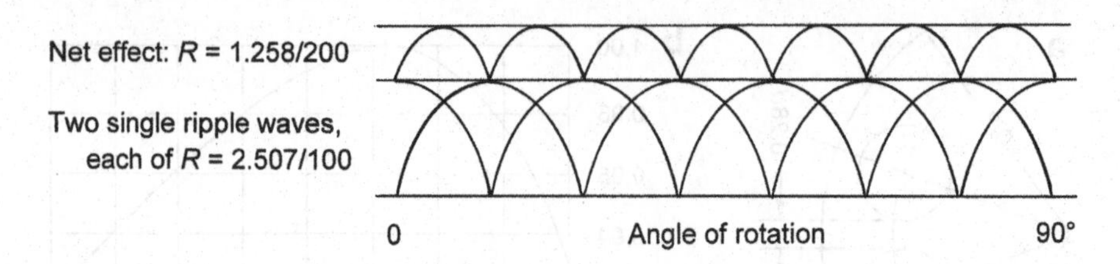

Fig. 12.8 Reducing the ripple to one-fourth its value by angle phasing two seven-piston units

This is about one-fourth the ripple of one unit. However, the original values of ripple do still exist in the torque and discharge flow rate of the individual pumps, or in the torque and the fluid inflow rate of the individual motors.

12.5.1 Phasing Back-To-Back Units

Back-to-back versions of hydraulic pumps and motors of the swashplate type have been suggested long ago with an aim to achieve a balanced axial-reaction design. Such units are still being manufactured to date according to the same basic configuration first suggested by North et al. (1968). They have in common a number of through bores in the cylinder block to receive pistons with slippers in both ends, while fluid porting is in the cylinder middle, in radial direction. However, these designs do not lend themselves to angle phasing of the two swashplates to reduce the ripple.

Recently, Friedrichsen et al. (2019) suggested a pump arrangement comprising two individual units arranged back-to-back with their swashplates fixed in phase and their cylinder blocks driven by a spline shaft such as to be assembled with an angular shift for reducing ripple. (The spline shaft was suggested with 18 teeth for nine-piston units, such that the phase shift would be 20°, which is not in accordance with the teachings in Sect. 12.5 that dictate a phase shift of only 10° in this case).

A rather new concept for the back-to-back arrangement of hydraulic pumps and motors is the floating-cup design. This concept was introduced by a number of inventors and manufacturers since the beginning of this millennium, with minute differences in detail and geometric proportions from the typical schematic shown in Fig. 12.9. The unit belongs to the bent-axis category. It features two arrays of pistons arranged back-to-back in a shaft flange. The pistons have spherical-segment heads that operate inside the cylindrical bores of cups, with or without piston rings. Each set of cups is seated onto a barrel plate, and means for retention (not shown) should be provided, preferably on the inside shoulder for maximum packaging density of the set of cups. Each barrel plate is also driven by the shaft by a pin–slot pair, supported on a spherical-segment of the shaft through a supporting ring, and pushed outwards by a wave spring (not shown) onto a wedge-shaped valve plate that contains the kidney-shaped fluid-porting slots as usual. Each valve plate sets

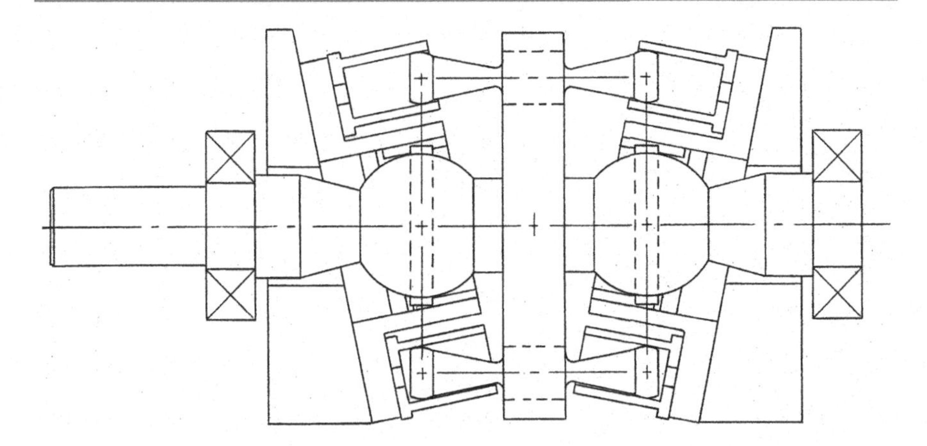

Fig. 12.9 Schematic of a hydraulic floating-cup unit

the location of the top and bottom dead centers for its pump/motor side independently. Therefore, the port timing of the two units can be angle phased to minimize the ripple.

An experimental verification of the effectiveness of angle phasing of the two units of a floating-cup pump was conducted by Achten (2004). The unit under consideration has 2 × 12 pistons which, when phase shifted by half a pitch (15°), will have the ripple reduced from (1 − cos 180°/12) = 3.41% to (1 − cos 180°/24) = 0.855%. Not mentioned was the fact that a 2 × 11-piston pump with the valve plates made out of phase by a quarter-pitch (8.2°) would reduce the ripple from 1.02 to 0.255%; almost to vanish.

References

Achten PAJ (2004) Power density of the floating cup axial piston principle. Paper presented at the ASME 2004 International Mechanical Engineering congress and exhibition, Anaheim, California, 13–19 November. https://doi.org/10.1115/IMECE2004-59006

Basstein AFH, Groenemans H (2018) A wind turbine. WO Patent 2018/078104, 3 May 2018

Dahlhaug OG (2013) Wind turbine with hydraulic swivel. U.S. Patent 8,405,238, 26 March 2013

De Vries E (2016) A wind turbine. EP Patent 3 096 007, 23 November 2016

Friedrichsen W, Andersen SK, Martensen L (2019) Pump arrangement. U.S. Patent 10,495,074, 3 December 2019

Janney R (1909) Variable speed transmission device. U.S. Patent 924,787, 15 June 1909

Lezard CV (1964) Hydraulic variable-speed drives. Engineers' Digest Survey No.18. Engineers' Digest, London

North JD, Millard DJ (1968) Hydraulic apparatus. U.S. Patent 3,407,745, 29 October 1968

Energy Retrieval, Storage, and Release 13

The two basic configurations of hydraulic kinetic energy retrieval systems (KERS) are described; those based on bidirectionally controlled, over-center variable-displacement pump/motor units, and on unidirectionally controlled ones. Over-center pump/motor units may be of the swashplate type, but the higher power density of the bent-axis types makes them more suitable. The (unspoken of) problems pertaining to the variable-displacement bent-axis units with tapered-feature synchronization of the cylinder block with the drive shaft are highlighted. The state of the art of variable-displacement units is presented and a design configuration of an over-venter pump/motor unit is accordingly herein suggested. Bladder accumulators used for energy storage are described and a worked example on their operation is given. It will also be shown how the performance density of KERS based on mechanical springs is unacceptably low.

13.1 Kinetic Energy Retrieval Systems

Kinetic energy retrieval systems (KERS) are used to convert the kinetic energy of a high-angular-momentum machine or a large-linear-momentum vehicle into a form of energy to be stored, for later release or deployment as work to re-accelerate the machine or vehicle. Energy is thus prevented from being dissipated into useless heat upon braking, and energy required to re-accelerate will be readily available from the storage. The retrieved energy could be stored either as kinetic energy in flywheels, pressure/potential energy in hydraulic accumulators, or electric energy in batteries or supercapacitors.

KERS necessitate the use of a variable-transmission-ratio module between the machine or vehicle and the energy storage system. In purely mechanical KERS a CVT is used with a flywheel, starting with a high-reduction ratio and controlled down to a low reduction toward the end of the storage process, then from a low step-up ratio to a high step-up ratio by the end of the retrieval process.

H. A. Arafa, *Design for Durability and Performance Density*,
https://doi.org/10.1007/978-3-030-56816-0_13

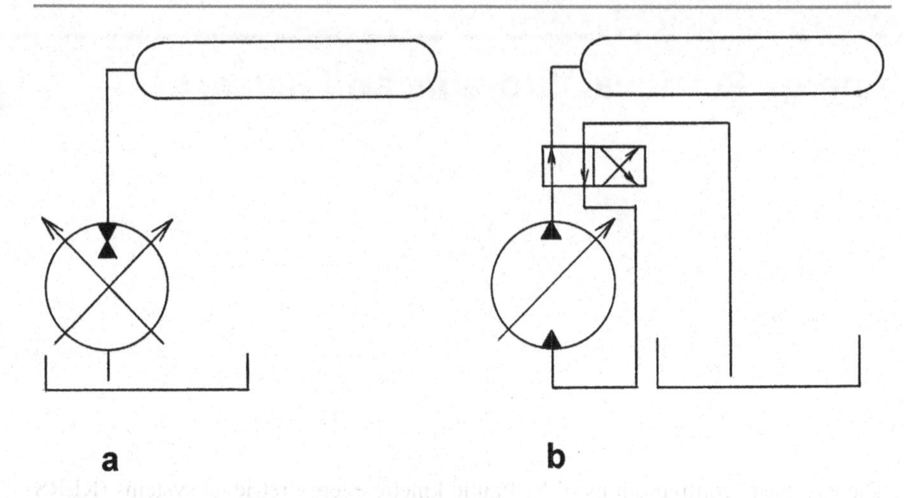

Fig. 13.1 Hydraulic kinetic energy retrieval systems, **a** with an over-center pump/motor, **b** with a variable-displacement pump/motor and a 4/2 way valve

Hydraulic KERS could be of either of two basic configurations, both in open-circuit, where a fluid reservoir and a hydraulic accumulator of ample capacities are necessary.

1. A bidirectionally controlled variable-displacement pump/motor unit (an over-center unit) is used; the fluid pumping direction being opposite the motoring direction, with one and the same direction of rotation as shown in Fig. 13.1a. The unit is so controlled during the storage and the retrieval cycles as to give best performance. The sequence in energy retrieval and release is discussed in Sect. 13.2, and a necessary account of over-center pump/motor units follows.
2. A unidirectionally controlled variable-displacement pump/motor unit is used in one and the same direction of rotation together with a 4/2 way valve between the reservoir, pump/motor unit, and the hydraulic accumulator, to switch between the two flow directions, every time, as shown in Fig. 13.1b.

In either case, a bent-axis pump/motor is preferred to the swashplate type for its higher energy efficiency. A 2/2 way valve should also be installed right before the hydraulic accumulator to seal it off when extended storage times are anticipated.

13.2 Sequence in Energy Retrieval and Release

Figure 13.2 depicts the sequence of states of a hydraulic KERS with an over-center unit in a descriptive way during energy recovery—in vehicle deceleration—and energy release—in acceleration. Four variable quantities are involved:

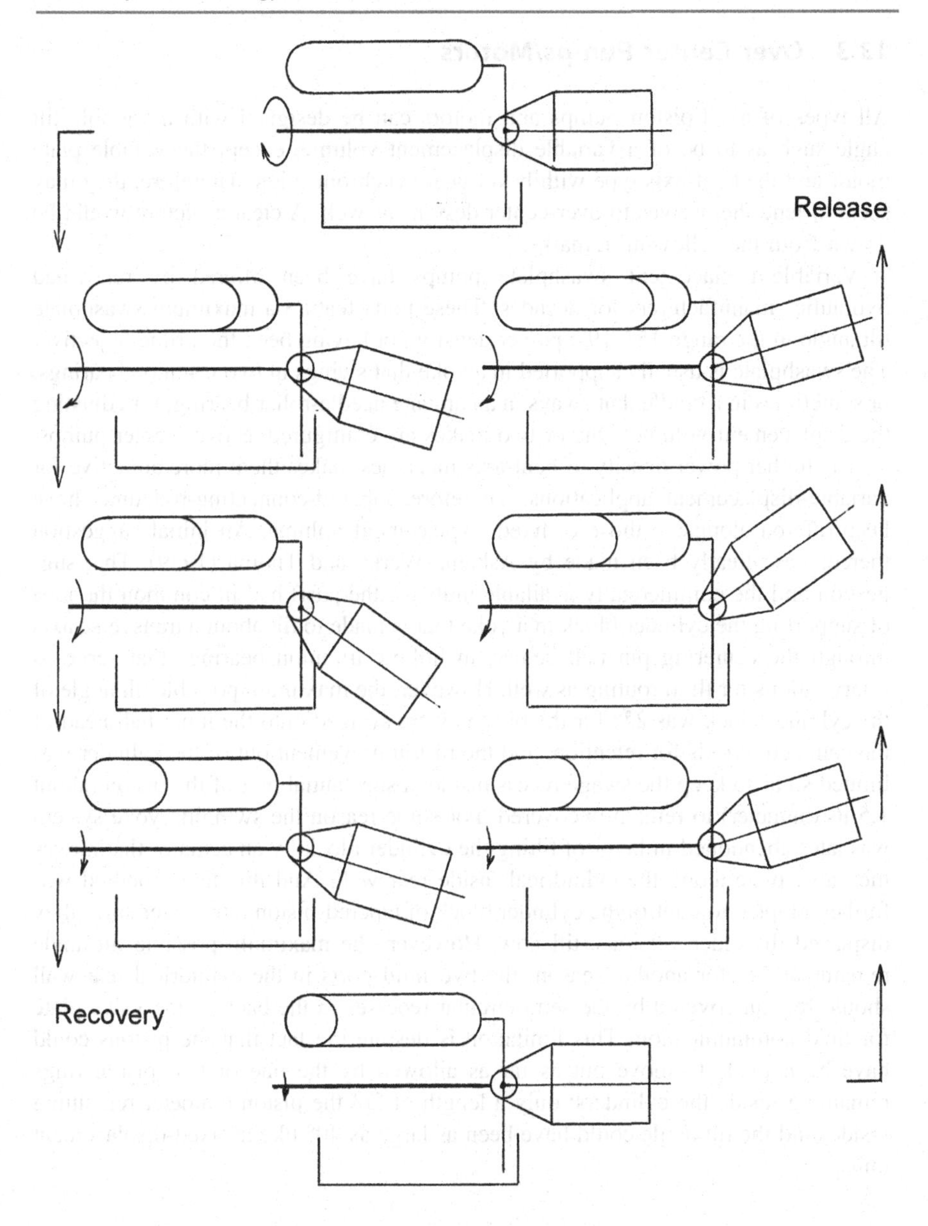

Fig. 13.2 Sequence of the states of a hydraulic kinetic energy retrieval system

1. Speed of shaft rotation, qualitatively indicated by the length of a curved arrow.
2. Tilt angle of the over-center unit, directly indicated in magnitude and direction.
3. Stored fluid volume in the accumulator, indicated by the bladder position (see Sect. 13.4.1).
4. Pressure of the stored fluid, also qualitatively indicated by the bladder position.

13.3 Over-Center Pumps/Motors

All types of axial-piston pumps and motors can be designed with a variable tilt angle such as to be of a variable displacement volume, except the wobble plate motor and the bent-axis type with bevel-gear synchronization. Therefore, they may seem to lend themselves to over-center designs as well. A clearer picture would be drawn from the following remarks.

Variable-displacement swashplate pumps have been offered by renowned hydraulics manufacturers for decades. These units feature a maximum swashplate tilt angle in the range 15°–19°; power density not having been the prime objective. The swashplate is usually supported in a yoke that swings in two trunnion bearings, or sometimes in a cradle that sways in an arcuate needle-roller bearing, for adjusting the displacement volume. One or two makes are configured as over-center pumps.

The higher power density of bent-axis machines makes them more attractive for variable-displacement applications. Therefore, tapered-connecting-rod units have been offered alongside those of fixed-displacement volume. An initial suggestion thereof has already been made by Askania-Werke and Thoma (1939). That suggestion and the commercially available units (of the past) had in common the idea of supporting the cylinder block in a yoke that is made to tilt about a transverse axis through the centering pin ball center, in hollow trunnion bearings that serve as rotary unions for fluid routing as well. However, the maximum possible tilt angle of the cylinder block was 27° for the pistons were swaged onto the inner ball head of the connecting rods for retention, and the piston movement out of the cylinder was limited so as to keep the swage recess inside; a substantial part of the piston (about 1.5 its diameter) to remain uncovered. For some reason the swinging yoke system was later abandoned in favor of tilting the cylinder block by an actuator that moves the valve plate along the cylindrical inside rear wall. And this latter method was further adopted to control the cylinder block of tapered-piston units, ever since they displaced the older versions, till now. However, the maximum possible tilt angle remains at 27° for another reason; the two fluid ports in the cylindrical rear wall should remain covered by the semi-circular recesses in the back of the valve plate for fluid communication. This limitation is despite the fact that the pistons could have been made to move out as far as allowed by the one or two piston rings remaining inside the cylinders; only a length of 2/3 the piston diameter remaining inside, and the tilt angle could have been as large as 40° like in fixed-displacement units.

13.3.1 Kinematics of Synchronization

If bent-axis units with tapered-feature synchronization are designed as variable-displacement ones as outlined in Sect. 13.3, then a *deeper* look is due to the kinematics (see also Sect. 11.2.2). Figure 13.3 shows the positions of the orbiting tapered features within the nine cylinder bores for the full tilt angle of

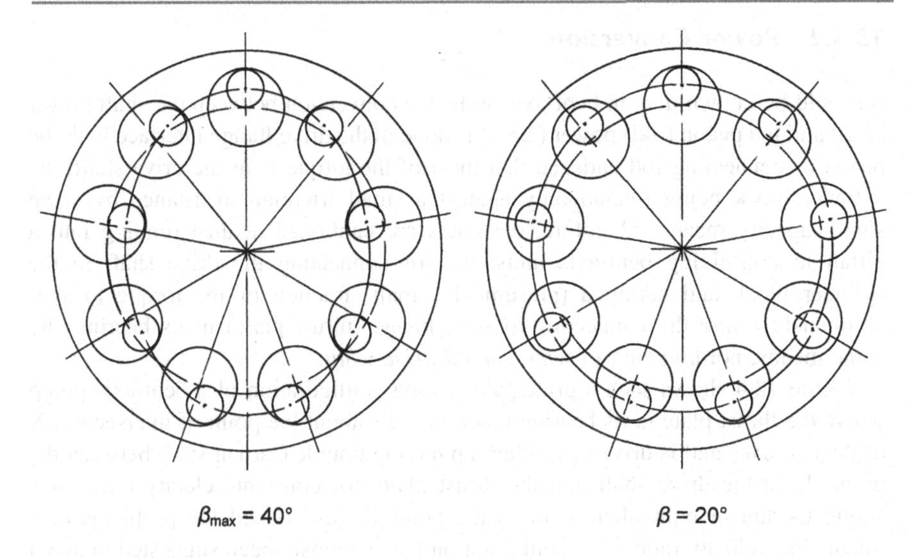

Fig. 13.3 Positions of the small ends of the tapered-pistons orbiting within the cylinder bores for cylinder block tilt angles $\beta_{max} = 40°$ and $\beta = 20°$

$\beta_{max} = 40°$ (just as in Fig. 11.3) alongside the positions when the cylinder block swings back, about the same vertical diameter, to $\beta = 20°$ (assuming synchronism being maintained by separate means). The latter is a condition that so far nobody felt like revealing; all the tapered features lose contact with the bores, except the one at the vertex of the elliptical trajectory. This one feature will momentarily assume the cylinder block driving function if a small clearance is provided as explained in Sect. 11.2.2, with some lag and at a very large pressure angle though. (The straight-line contact is maintained in this kinematic geometry, along a common generatrix). After a few degrees of rotation the same is taken over by a diametrically opposite piston, and so forth. The cylinder block will thus be driven from two opposite points in a fast alternation. This condition worsens; the lag increases much as the tilt angle is further reduced.

However, variable-displacement units of this design have been in successful operation in unidirectional control for decades, and they still are, because of the rather small frictional resistance torque on the cylinder block. But the large pressure angle at the drive contacts makes them very sensitive to dynamic torque, so that acceleration and deceleration of the drive shaft should be largely avoided. Over-center units should not be designed with tapered-feature synchronization because the excessive angle of lag around the null position would show-up in both directions, due to the repeated null crossing from the pumping to the motoring mode and back or the reversal of the rotation direction, or both, and the timing of the valve plate would not always be correct.

13.3.2 Power Conversion

It should be known that, in bent-axis units, the conversion between the shaft power ($T.\omega$) and the piston (rod) power ($\Sigma F.v$) is done in the drive flange interface with the piston or connecting rod ends, so that most of the torque is in the drive shaft; the cylinder block being rotated only against a small frictional resistance by some small-capacity means. Therefore, designers are cautioned against running into a pitfall in conceiving bent-axis units; that of connecting the drive shaft to the cylinder block and taking a (presumably small) fraction of the torque to synchronously *rotate* the connecting-rod-end flange (thrust plate) in its bearings by some means; not knowing of *what should drive what.*

Figure 13.4 shows such a principally wrong configuration of a bent-axis pump where the thrust plate in its bearing block swivels about the point of intersection X of the two axes and is driven by either a plunging double Cardan shaft between the points U in the drive shaft and the thrust plate, for constant-velocity ratio, or a single U-joint or equivalent at the same point X, just to add the problems of a fluctuating velocity ratio. This configuration has obviously been suggested to avoid the rather long-winded routs of oil when the cylinder block is the swiveling member. A recent example of this design pitfall with an over-center pump/motor (intended for a KERS) is that in the patent by Hoover (2013), which pitfall was repeated after a three-quarter century from a similar suggestion by Wegerdt (1942).

13.3.3 State of the Art of Variable-Displacement Units

Most recent variable-displacement, axial-piston pumps and motors feature spherical-head pistons and synchronization by a double tripod shaft for unidirectional cylinder-block tilting by the valve plate sliding on a semi-cylindrical rear wall of the casing. These pumps/motors are available from a couple of world-renowned

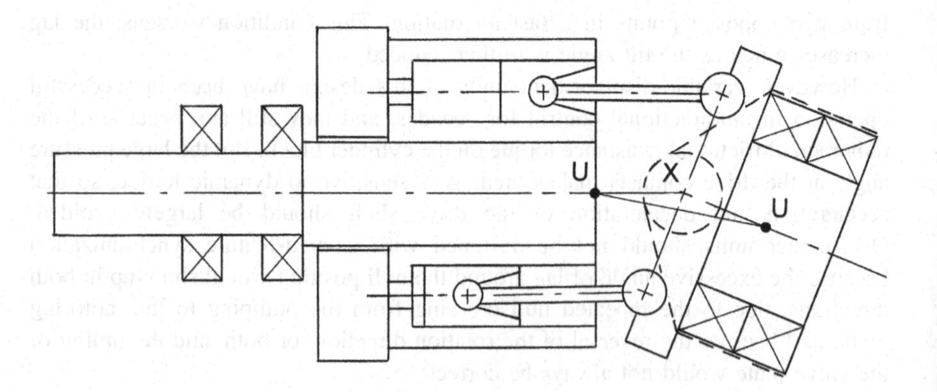

Fig. 13.4 Bent-axis pump where the thrust plate is driven from the pump shaft

hydraulics manufacturers. The maximum possible tilt angle is 35° for the coverage of the fluid ports as mentioned in Sect. 13.3. Power density is therefore at its best, and many of the above cited problems are nonexistent. For these advantages to be imported into over-center units, only the swinging and fluid porting should be changed back into the old proven but longer-winded yoke and rotary union system.

A herein suggested schematic sectional drawing in two views is given in Fig. 13.5 of an over-center pump/motor according to the ideas just outlined, which should make it KERS-worthy. The design is of a nine-piston unit with $\beta_{max} = \pm 35°$. The same idea has been suggested by Ryken et al. (2001) for $\beta_{max} = \pm 45°$, with more slender pistons. However, this is a domain where the two attributes of high power density of the unit and durability of the tripod joints could be seen as competing design objectives that require striking a trade-off.

13.4 Hydraulic Accumulators

Hydraulic accumulators are hydro-pneumatic energy storing devices that are connected in parallel with the high-pressure line in a hydraulic system to store/release energy in accordance with an increase/decrease of the fluid pressure in that line. The higher the pressure the faster will be the response in the energy release mode. The energy storage and release function is needed for the following.

1. Saving pump driving power in intermittently operating systems.
2. Coping with temporary peak flow demands.
3. Mitigating vibrations due to pump or motor ripple.
4. KERS.

An energy storing/release device is characterized by the maximum amount of energy it could store per unit mass; the specific energy, and the maximum rate of releasing that energy per unit mass; the specific power. The latter is customarily referred to as the power density. The maximum power is limited to a value that maintains the integrity and an adequate durability of the storing device, and is attained at the first instant of release. Dividing the two figures of merit gives the time taken to deplete the stored energy at its initial rate, but the actual time is (much) longer since the energy release rate decays. These characteristics, of one device, can be represented by a point on the so-called Ragone chart, on log–log scales to accommodate data entries of widely different devices, and to make the isochronous lines of unity slope. Figure 13.6 shows a Ragone chart, empty though, since the data points vary much between the different sources and no liability could be assumed for correctness. The diagram is also called a bubble chart, since the data of any one device type lies scattered within a zone, or a bubble.

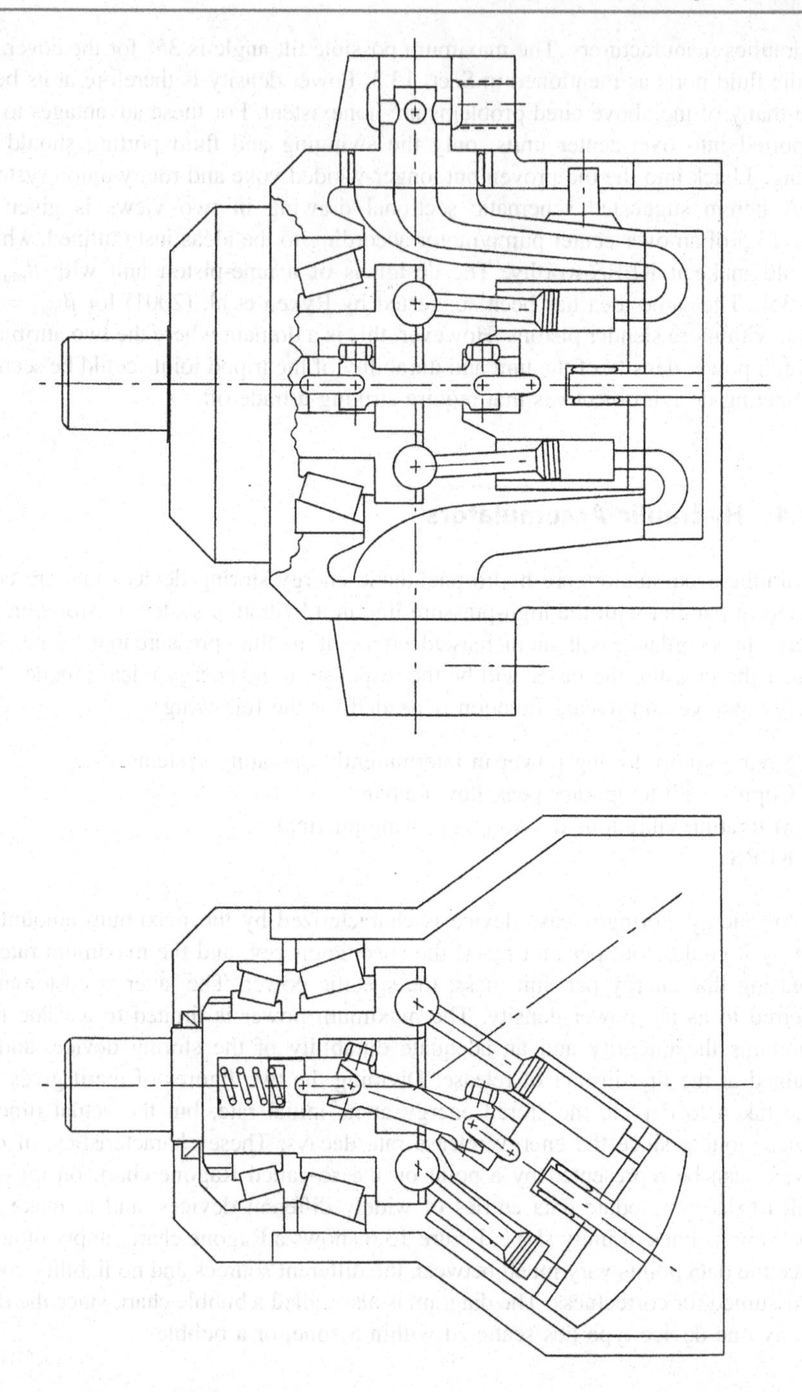

Fig. 13.5 Suggested schematic of an over-center pump/motor of high power density, shown in two orthographic sectional views

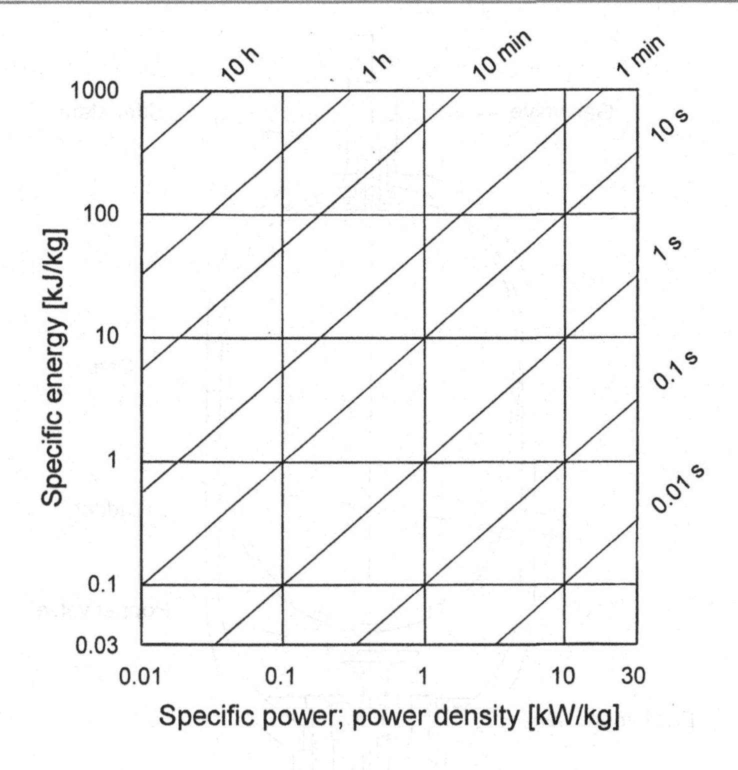

Fig. 13.6 Empty ragone chart

The Ragone chart was initially introduced to compare electric energy storage devices such as batteries and capacitors of all the different technologies, types, and sizes, but then the same scale ranges could include flywheels, hydraulic accumulators, and even springs.

In general, these charts reveal that composite flywheels are superior to steel flywheels, and composite accumulators are superior to steel accumulators, both hydraulic accumulators being of an order of magnitude higher specific power than the flywheels. Therefore, hydraulic KERS of best performance density are those with composite accumulators; they have been specifically developed for and successfully used in medium and heavy commercial vehicles.

13.4.1 Bladder Accumulators

The type of hydraulic accumulators of highest energy storage capacity and smallest hysteresis is the bladder type, shown schematically in Fig. 13.7. A pressure vessel (shell) is either filament-wound in case of composite accumulators, or forged from high-tensile steel in case of steel accumulators, as a cylindrical skirt with two

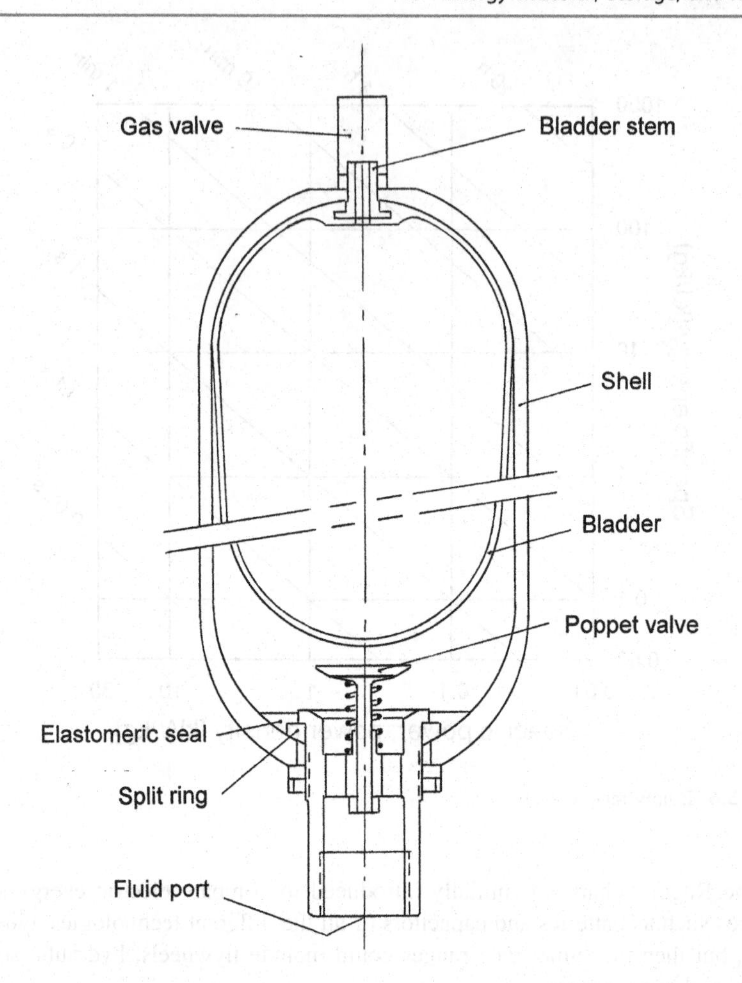

Fig. 13.7 Bladder accumulator

semi-spherical domes. An elastomeric bladder is provided with a bladder stem and inserted into the pressure vessel through the larger opening at the bottom. The stem is secured to the smaller opening at the top, and a gas charging valve is fitted therein. A poppet valve cartridge is assembled in the fluid port at the bottom; the poppet valve being held open by a spring and serves to prevent the bladder from being extruded out of the fluid port under the gas pressure upon shutdown of the hydraulic system.

The accumulator is specified by its nominal volume V_{nom} of its empty shell, as depicted in condition 1 in Fig. 13.8. The bladder is then inflated with pure nitrogen gas at a filling pressure p_0 (chosen according to the intended application) to fill the shell, shutting the poppet valve, condition 2. Subtracting the volume of rubber, the initial inside volume of gas V_{in} will amount to about $0.95V_{nom}$. In operation—with a

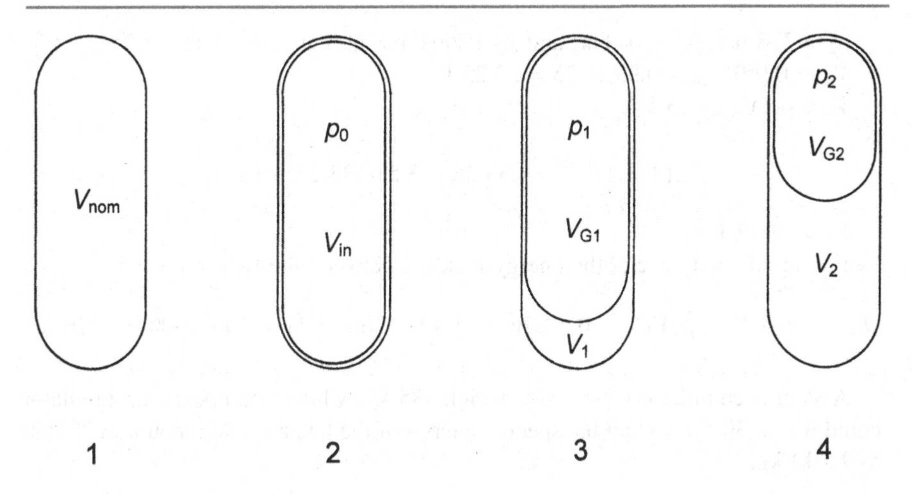

Fig. 13.8 Bladder accumulator conditions, from an empty shell to under maximum system pressure

system pressure $p_S > p_0$—the bladder divides V_{in} into a variable gas volume V_G inside the bladder and a variable fluid volume V outside it, the two volumes being complementary. Condition 3 depicts the minimum system pressure p_1 that should not drop below $1.1p_0$ in order to maintain a minimum amount of fluid $V_1 \approx 0.1V_{nom}$ for the bladder not to hit the poppet valve by the end of every cycle. Condition 4 is when the accumulator is under maximum system pressure p_2, which should be kept equal or below $4p_0$, setting an upper limit on the fluid volume V_2 for the bladder not to be subjected to unduly large-amplitude fatigue cycles, as it deforms into a pear shape and back. The maximum admissible pressure fluctuation range $p_2 - p_1$ is usually 200 bar, according to practice with the operation of bladder accumulators.

Hydraulic accumulators are intended for storing/releasing energy in short time durations; not exceeding a few seconds. The gas behavior will then be adiabatic, so that cyclic operation between two fluid pressures p_1 and p_2 and two fluid volumes V_1 and V_2 follows the relationship

$$p_2/p_1 = (V_{G1}/V_{G2})^\gamma = [(V_{in} - V_1)/(V_{in} - V_2)]^\gamma$$

where γ is the specific heat ratio of nitrogen (air), $\gamma = 1.4$.

13.4.2 Worked Example

A bladder accumulator of nominal volume $V_{nom} = 35$ L is to be used in its maximum allowable energy storage/release mode. Calculate the pressures, volumes, and amount of energy handled.

$p_2/p_1 = 4/1.1$ while $(p_2 - p_1) = 200$ bar, therefore.

$p_2 = 276$ bar, $p_1 = 76$ bar, and $p_0 = 76/1.1 = 69$ bar.
$V_{in} = 0.95V_{nom} = 0.95 \times 35 = 33.25$ L.
$V_1 = 0.1V_{nom} = 3.5$ L

$$(4/1.1)^{1/1.4} = (33.25 - 3.5)/(33.25 - V_2)$$

$V_2 = 21.42$ L.
For an adiabatic cycle, the energy handled between the two states is

$$E \approx 0.4(p_2V_2 - p_1V_1) = 0.4(276 \times 21.42 - 76 \times 3.5) \times 100/1000 = 226 \text{ kJ}$$

A steel accumulator of this size weighs 85 kg, while a composite accumulator could weigh 30 kg, so that the specific energy of the latter would amount to 226/30 $\approx$ 7.5 kJ/kg.

13.5 Energy Storage in Mechanical Springs

By virtue of the high strength properties of spring steels and their finite modulus of elasticity, springs store a reasonable amount of elastic or strain energy when deflected. The energy could be stored for indefinite time duration and released for re-use at arbitrary rates without internal losses. The energy storage capacity per unit material volume is called the energy density, and that per unit mass is called the specific energy. It varies according to the allowable maximum stressing level, the nature of stressing the material elements (tensile, bending, or torsional shear), the stress distribution pattern in the material section(s), and the distribution of loading along the spring.

Full utilization of the spring material for energy storage is obtained when every material element is stressed up to the same allowable tensile stress σ. This is obtained only with a material of an extruded geometry under tension/compression. When subjected to an increasing force up to F, under which a linear elastic deformation δ is reached, the stored energy density will be $F\delta/(2V) = \sigma^2/(2E)$. This parameter is expressed in units of stress, equivalent to energy per unit volume; J/m^3. The quantity $\sigma^2/(2E)$ is termed the specific potential energy of the spring material. Similarly, in a hypothetical case where every material element is stressed up to the same allowable shear stress τ, the specific potential energy of the spring material will be $\tau^2/(2G)$. The same values are obtained if $\tau = 0.62\sigma$.

An example is due to show the order of magnitude of storable elastic energy in springs. Taking a spring steel that is loadable to $\sigma = 800$ MPa, with $E = 207$ GPa, then its specific potential energy will amount to 1.546×10^6 J/m^3. With a steel density of 7,850 kg/m^3 the specific energy will be about 200 J/kg. On the other hand, one kilogram of steel acquires a kinetic energy of 200 J when moving at 72 km/h, so that a spring (wire), pre-stressed to its maximum, would accelerate to 72 km/h if it could appropriately be triggered to move in a straight direction.

Therefore, a car of any mass, moving at 72 km/h, would require springs of its same mass to store its kinetic energy when decelerated to zero velocity. Kinetic energy retrieval systems with springs are thus not feasible at all for automotive driveline applications; their performance density is unacceptably low.

The low-deflection high-load characteristics of a straight wire as in the example above are, however, not practical. They necessitate high-ratio amplifying devices such as levers or gearing, which may be associated with energy losses. Therefore, springs are made in other configurations, as well-known. The stress distribution over the material section will differ from one type to another, and this could be expressed by a form factor that relates the actual energy density to $\sigma^2/(2E)$; smaller values of the energy density being obtained for the different types of springs.

References

Askania-Werke, Thoma H (1939) Hubkolbengetriebe. CH Patent 206998, 15 September 1939

Hoover DV (2013) Bent axis variable delivery inline drive axial piston pump and/or motor. WO Patent 2013/116430, 8 August 2013

Ryken JD, Dirks DD, Kardell D, et al (2001) Ball joint for servo piston actuation in a bent axis hydraulic unit. U.S. Patent 6,257,119, 10 July 2001

Wegerdt F (1942) Variable speed hydraulic power transmission device. U.S. Patent 2,297,518, 29 September 1942

14 Multi-Attribute Designs

Design features are those readily visible geometric and numeral details of the design, as well as other *invisible* particulars such as freedom of kinematic over-constraints in assemblies, or freedom of power recirculation in planetary gearing. The latter two may also be called attributes, just like durability and power density are. Designing with the objective of multiple attributes is the process of creating a product that excels in integrating more than just one desired attribute, by adopting one or more ingenious mechanical design ideas. Such a product should then, after being put into service, prove to belong to the so-called surviving or well-proven designs. Examples will be given of some multi-attribute designs from the technologies of fluid power, double-helical gearing, and curved-tooth gears. The latter have been subject of invention and investigation for over a century and a half without nearly approaching viable generating kinematics for the types of choice that promise optimum operational and performance characteristics.

14.1 Well-Proven Radial-Piston Motor

A hydraulic radial-piston motor of well-proven design is shown in Fig. 14.1, in which most of the effort is transmitted between the casing and the crankshaft directly by means of fluid columns, rather than the more usual pistons and connecting rods. The fluid column is contained inside a two-piece, open-ended telescopic cylinder of which the end faces are spherical-segment ring areas that seal against corresponding surfaces in the cylinder heads and the spherical-segment crankpin. The remainder of the effort is transmitted by mechanical contact of these surfaces, and equals the fluid pressure times the (small) annular projected area

H. A. Arafa, *Design for Durability and Performance Density*,
https://doi.org/10.1007/978-3-030-56816-0_14

between the cylindrical interface and the inside diameter of the sealing lip of the outer cylinder, which should be about equal the same with the inner cylinder. Therefore, the fluid pressure drives the crankshaft and increases the sealing force proportionally; on-demand, satisfying the design principle of self-reinforcement. The telescopic cylinders are lightly spring loaded to maintain contact during the return stroke, also when the motor is not in operation. This design also satisfies the principles of minimum interface sliding volume per cycle because of the minimal direct contact surface areas involved, and of homogeneous-wear interface by nature of the spherical contacts. More importantly, this oscillating cylinder mechanism is of zero pressure angle; no side thrust exists between the cylinder pairs. Finally, this is a quasi-exact-constraint design, which performs jam-free regardless of any elastic deformations under load or thermal deformations. Therefore, hydraulic radial-piston motors according to the present design enjoy both attributes of high durability and performance density. High efficiency is also a subordinate attribute; the best-efficiency operating zone could exceed 90% energy efficiency. The larger sizes of this motor are referred to as low-speed high-torque motors.

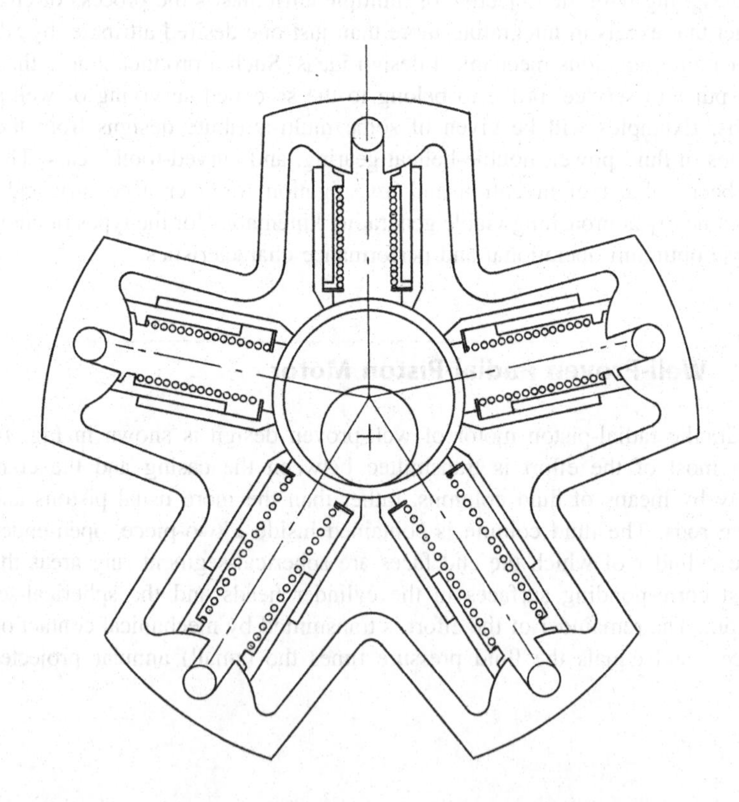

Fig. 14.1 Hydraulic radial-piston motor with five fluid-column telescopic pistons

14.2 High Power Density Bent-Axis Unit

Hydraulic axial-piston pumps or motors of the bent-axis type provide the piston stroking in their cylinders by virtue of the tilt angle between the drive shaft and the cylinder block axes. These units come in several versions regarding the method of synchronizing the cylinder block with the drive shaft and, accordingly, the piston geometry. In the particular version shown in Fig. 14.2 the cylinder block is driven by a pair of equal, narrow-face bevel gears from the drive shaft. This version is a good example of multi-attribute designs, since it integrates several features and satisfies several design principles, which are summarized in the following.

1. The (wobbling) piston inclination to the cylinder axis is very small, typically < 4°, which satisfies the principle of minimum pressure angle for minimum side thrust, allowing operation at high pressures, hence high power density.
2. The tilt angle is typically made 40° and as large as 45° in some brands, giving highest possible displacement volume, hence highest power density.
3. The piston stroking and the cylinder block synchronizing are allocated to separate function carriers, satisfying the principle of separate allocation of functions.
4. This design is kinematically exactly constrained, posing no particularly stringent requirements on the dimensional and geometric accuracy of the inter-interface bodies (see Sect. 7.3), and resulting in markedly good durability.

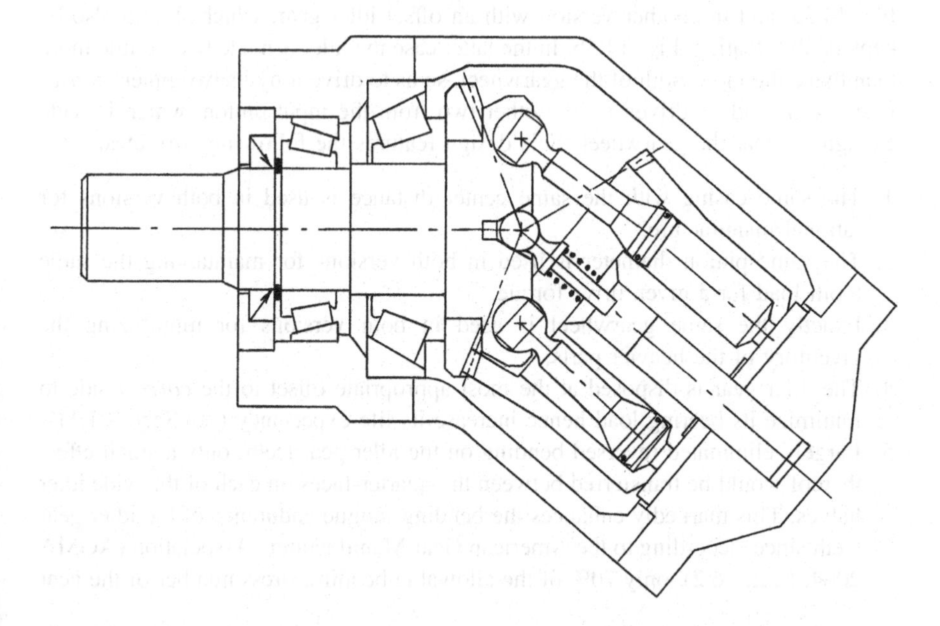

Fig. 14.2 Hydraulic bent-axis unit with bevel-gear synchronization of the cylinder block

5. One additional attribute of this design is that it satisfies the requirements to qualify as a servomotor; the bevel-gear angular backlash between the drive flange and the cylinder block is almost zero, and at a large radius, thus the operation smoothly transits between the two senses of rotation. This is not the case with any of the other synchronization means adopted for such units. The high power/torque density, which also implies a high-torque-to-inertia ratio of the rotating parts, renders the servomotor fast response characteristics.

14.3 Turboprop Gearbox for Both Directions of Rotation

Two- and four-engine aircraft are sometimes designed with counter-rotating propellers for cancelling out their gyroscopic effects, and for better propulsion aerodynamics in the latter case when each two counter-rotating propellers are on one wing. Turboshaft engines are available in only one direction of rotation; for rational manufacture and minimal parts inventory. Therefore, the speed reducing gearboxes between the engine and the propeller are needed in two alternate quasi-identical versions; for either direction of rotation. In either case the input stage is typically followed by a planetary reduction stage.

An input-stage reduction gear set that is designed in direct mesh for opposite directions of input and output rotations or else with an idler for the same directions is shown in Fig. 14.3. It is a double-helical gearing configuration, in a version featuring an axially floating input pinion in direct mesh with the gearwheel, Fig. 14.3a, and in another version with an offset idler gear, which should also be kept axially floating, Fig. 14.3b. In the latter case the idler is made with a little more than twice the face width of the gearwheel, so as to drive it by its two inner *quarter* face widths and be driven by the other two from the input pinion, which is wide enough to clear the gearwheel. This design features the following attributes.

1. The same casing with the same center distance is used in both versions for rational manufacture.
2. The same pinion diameter is used in both versions for maintaining the same tooth load for a given drive torque.
3. Exactly the same gearwheel is used in both versions for minimizing the inventory of the heavier parts.
4. The idler gear is disposed at the most appropriate offset to the *correct* side to minimize its bearing load hence increase its life expectancy (see Sect. 2.1.1).
5. Largely eliminated reversed bending on the idler gear teeth; only a small effect thereof would be transferred between the quarter-faces on each of the wide idler halves. This markedly enhances the bending fatigue endurance of the idler gear teeth since, according to the American Gear Manufacturers Association (AGMA 2004, Sect. 16.2), only 70% of the allowable bending stress number of the gear

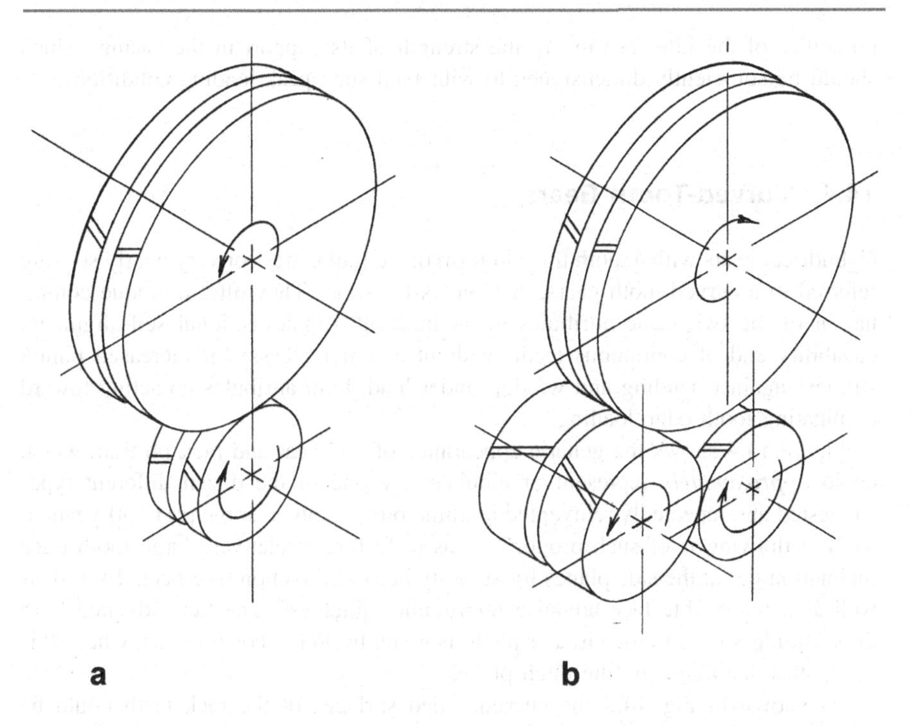

Fig. 14.3 Input stage of a turboprop gearbox, **a** direct mesh configuration for reversing the rotation direction, **b** with an offset idler gear for the same rotation direction

steel should be used for idlers "where the teeth are completely reverse loaded on every cycle." This 0.70 factor is called the *reverse loading factor*.

This idea is being implemented with a reduction ratio of about 2.1 in the first stage of the turboprop gearboxes of a recently introduced heavy transport aircraft with counter-rotating propellers, as depicted in a presentation by Andrei (2009). Attributes 2, 3, and 5 above should be compared with *purely coplanar* gearing in which a version with an idler implies decreasing the numbers of teeth of the pinion and gearwheel in the same ratio as the reduction ratio (if possible) for maintaining the same center distance while clearing the tooth tips, and in which the idler gear teeth are subjected to reversed bending.

Unidirectional loading in turboprop gearboxes is not always guaranteed; there could be so-called negative torque situations such as in case of *windmilling* where —after a flameout—the propeller tends to drive the engine to restart, or due to some other reasons. A gearbox with an idler would then be at a disadvantage since the idler will become at the improper location with its bearing loaded several times the value it is designed for (see Fig. 2.1). Negative torque situations are expected but for relatively short time durations so that the main concern will shift from the durability of the idler bearing to the strength of its support in the casing, which should be sufficiently dimensioned to withstand such peak loading conditions.

14.4 Curved-Tooth Gears

Cylindrical gears with a tooth trace in form of a circular arc—or very nearly so—are referred to a curved-tooth gears, or C-gears for short. They offer a unique combination of the two basic attributes of an inherent, omnidirectional self-alignment capability and of continuous teeth (without a center recess) for increased pinion stiffness against bending and windup under load, both attributes co-acting toward eliminating tooth edge loading.

Figure 14.4 shows the generic appearance of a C-gear and rack portion, which could *approximately* represent or emulate any one of the dozen different types suggested and repeatedly reinvented in numerous patents over the past 150 years or so. The drawing is of such proportions as to feature a relatively large tooth trace inclination (ψ) at the side planes for stability in axial direction (see Sect. 14.4.5), as well as a reasonable face advance for running quietness. The face advance, as a dimensionless ratio to the circular pitch, is given by $R(1 - \cos \psi)/(\pi m)$, where R is the tooth trace radius on the pitch plane.

As shown in Fig. 14.5 the curved, ruled surfaces of the rack teeth could be conceived as parallel slices or stripes cut out of either one of the following geometrical shapes, with minor differences that could not be perceived in the drawing.

1. A right circular cylindrical surface, cut inclined at the pressure angle (φ) to its base.
2. An oblique circular cylindrical surface inclined at the pressure angle, cut parallel to its base.
3. A right circular conical surface with a semi-cone angle equaling the pressure angle, cut transversely.

The convex flanks are produced on the outside and the concave flanks on the inside of these surfaces. In all three cases the flanks are to be generated by blades in rotating cutter heads in intermittent or single indexing processes to produce teeth that are symmetrical about the midplane. Further less relevant suggestions were of these surfaces as slices of curtate or looped trochoids of the same geometrical shapes mentioned, when generated in continuous indexing processes, as well as of much longer-winded geometry, sometimes inexact. A comprehensive survey on the geometry and machining kinematics of C-gears was presented by Arafa (2005). It should be clear that cutting C-racks or internal gears is not possible due to the circular motion of the tools.

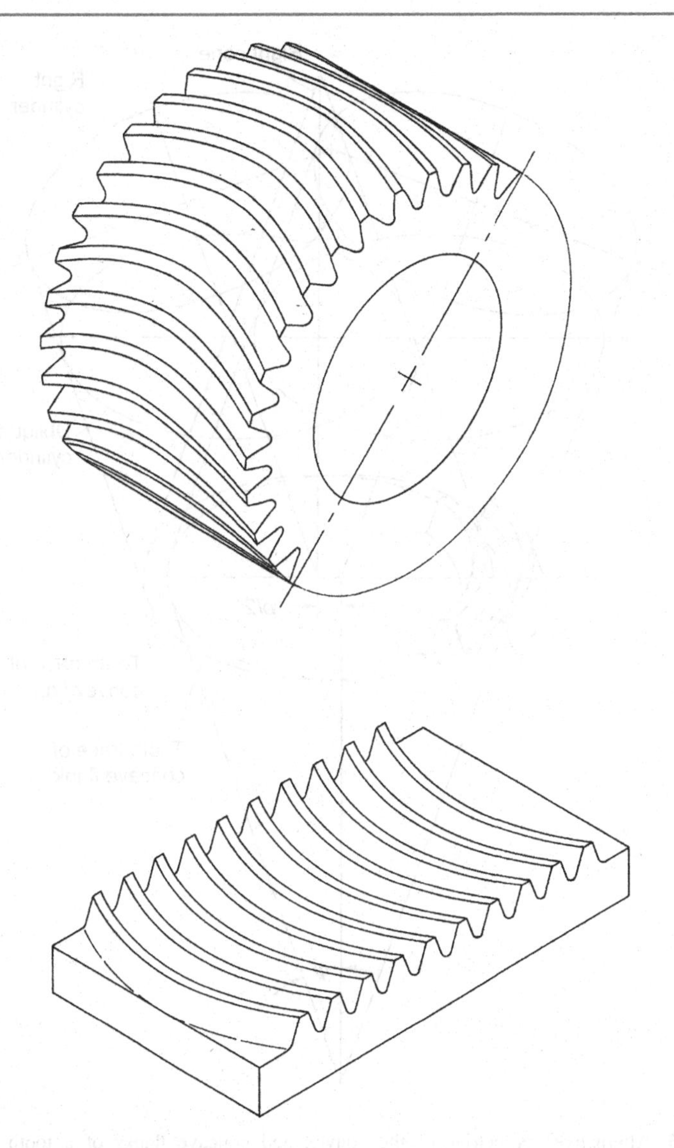

Number of gear teeth $N = 30$
Face-width-to-module ratio $f/m = 20.41$
Tooth-trace-radius-to-module ratio $R/m = 18.15$
Tooth trace inclination at the side planes $\psi = \arcsin(f/2R) = 34.21°$
Face advance = $R(1 - \cos\psi)/\pi m = 1.0$

Fig. 14.4 Generic appearance of a curved-tooth gear and rack, of such proportions as to be of a face advance of unity

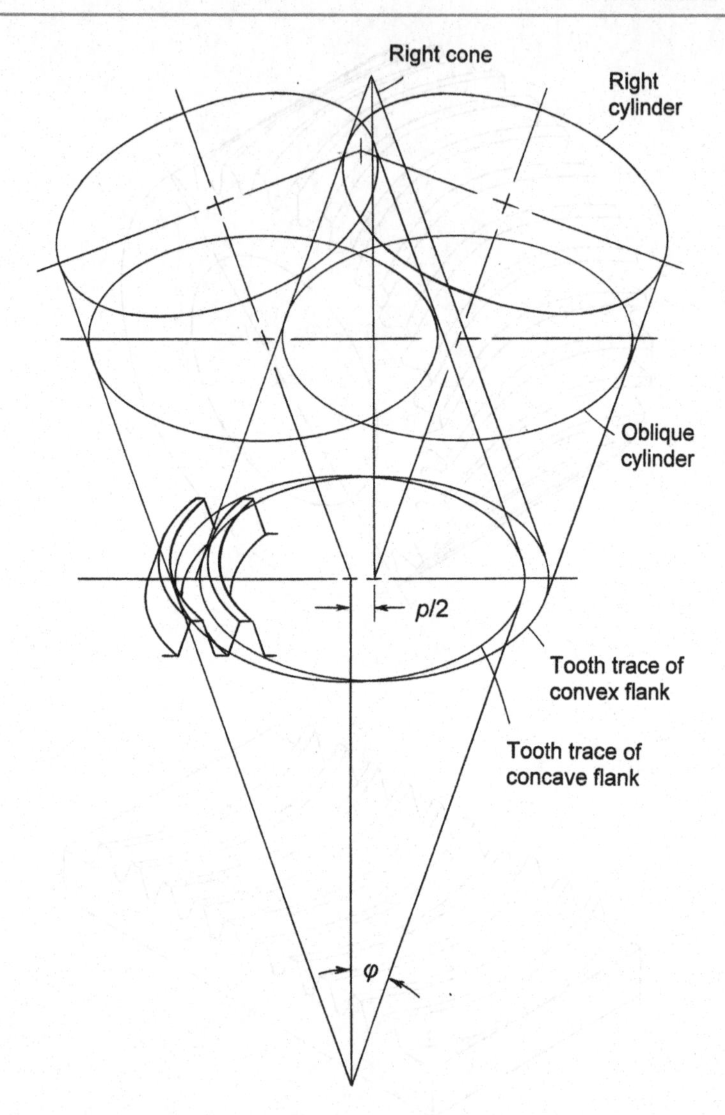

Fig. 14.5 Alternative geometries of the convex and concave flanks of a tooth space of a curved-tooth rack. The labeled geometrical shapes are of the convex tooth flank, the opposite ones for the concave tooth flank

14.4.1 Involute C-gears of Constant Pressure Angle

Despite the large variety of C-gears regarding the geometric details, only the two firstly mentioned types in Sect. 14.4 do cope best with the known stringent functional and operational requirements of gearing. This is by virtue of them having a true-involute profile in all transverse sections, with a constant transverse pressure angle throughout, and full-face line contact (see Sects. 14.4.4 and 14.4.5). They will also be insensitive to center distance changes. These attributes, in conjunction with the two basic ones are just what is needed for heavily loaded gearing, particularly for the input or the only power-splitting, step-up stage in wind turbine transmissions of multi-MW power rating, where the tooth load approaches a hundred tons. For such a configuration C-gearing with each pinion straddle-mounted in two (slightly) toroidal roller bearings, or even well designed journal bearings, would become the unbeatable solution in terms of the attributes of durability and torque and power density. However, it has not been possible to produce these types of C-gears in viable machining processes, without limitations or compromises. The present Chapter in addition to the two papers by Arafa et al. (2010a, 2010b) may encourage the invention of serious, patent-worthy kinematic embodiments for machining either type. It could be noted that renewed interest of world renowned rotorcraft manufacturers in C-gears (due to their self-aligning characteristics and the elimination of edge loading) was recently evidenced by Bittner (2018).

14.4.2 State of the Art

Industry may now be standing at the crossroads between acquaintance and satisfaction with available gear manufacturing machines, and a need for robust design actions on the problem of premature/unpredictable failure of heavily loaded gearing, primarily due to edge loading. It is logical that, on the one hand, engineers only design and specify gears for which cutting and finishing equipment exist and that, on the other hand, the machine tool industry will only manufacture the machines required for the types of gears specified by those engineers. Therefore, introducing a new type of gearing that needs other-than-conventional equipment would imply a paradigm shift in this discipline; it requires bold decision making based on undisputable technical/economic justification. The introduction of C-gears may thus be pioneered by a machine tool builder who plans to collaborate with a gearbox manufacturer for competing in the field of heavily loaded drive trains, particularly those of rotorcraft as well as horizontal-axis wind turbines.

14.4.3 Increased Torque Density by Curved-Tooth Gears

Increasing the torque and power density of rotorcraft transmissions has been and still is subject of extensive efforts, which are summarized in the following.

1. Engineering new steels for gears (and bearings) is considered to have matured; improvements today would be incremental. According to AGMA (2004, Table 3) the ceiling of the allowable contact stress number remains at 1,896 MPa (275 ksi) for the best available carburized and hardened Grade 3 steels. This value has been experimentally re-affirmed by Lewicki et al. (2008). Breakthroughs may or may not be imminent.

2. The adoption of split-power-path gearing configurations with as many as possible final-drive pinions to drive a large combining gear has proven to contribute to weight saving, more significantly in the larger helicopters.
3. The recent suggestions to eliminate the center recess in double-helical gearing; coming up with pseudo-herringbone pairs, but the teeth remain discontinuous.
4. The multiple, partly interlinked features of curved-tooth gearing are deemed to lead to higher torque density and improved reliability, Arafa et al. (2010b). The higher torque density is obtained through a more even load distribution across the face width; a closer-to-unity load distribution factor in the gear rating formulas. This is made possible by evading edge loading by virtue of the fundamental attribute of self-alignment of the curved teeth, in addition to the continuous teeth without a center recess for less pinion bowing and gear rim cupping. The weight saving due to the continuous teeth also contributes to a higher torque density. Higher reliability is clearly a direct outcome of the foregoing features.

14.4.4 Gears of Base-Surface Circular Tooth Trace

CC-gears This type is obtained when the rack tooth flanks are conceived as identical slices of a circular cylindrical curtain, cut in parallel at an angle ($90° - \varphi$) to the cylinder axis, and arranged alternately to expose their outside and inside surfaces as convex and concave flanks, respectively. The rack thus formed could then be what was meant by Böttcher (1913) in introducing his "better" conical-flank-rack type of C-gears, when referring to "previously known gears"; "While the flank surface of a straight involute rack tooth represents a plane, it becomes a section of a cylinder curtain in the known circular-arc-shaped curved tooth," yet citing no reference. Böttcher's gearing suffers the two drawbacks of a pressure angle that increases from the midplane toward the side planes, where also the transverse involute profile becomes distorted. But in a rack tooth of which the flanks are slices of a circular cylinder, all the transverse sections will be identical trapezoids with the pressure angle remaining constant over the face width. Gears conjugate to this rack will then feature teeth of the same true-involute transverse sections and a constant pressure angle throughout. Correct involute action will thus be obtained in all transverse planes of a pair of such gears. They will be designated CC-gears; for the circular-curved or cylindrical rack tooth flanks. The tooth trace is a circle scribed on the base cylinder unwrapped into the plane of action. The tooth trace on the rack pitch plane will be an ellipse of an aspect ratio sec φ with its major axis in the midplane, giving a slightly better face advance than a circular tooth trace.

The basic generating kinematics of CC-gears is shown in Fig. 14.6. These gears could simply be generated using two cylindrical cutters; blades arranged on cylindrical curtains with their straight cutting edges parallel to the cylinder axis, for inside-cutting the convex flanks and for outside-cutting the concave ones separately, in generating roll over the base cylinder. The involute generation proper is

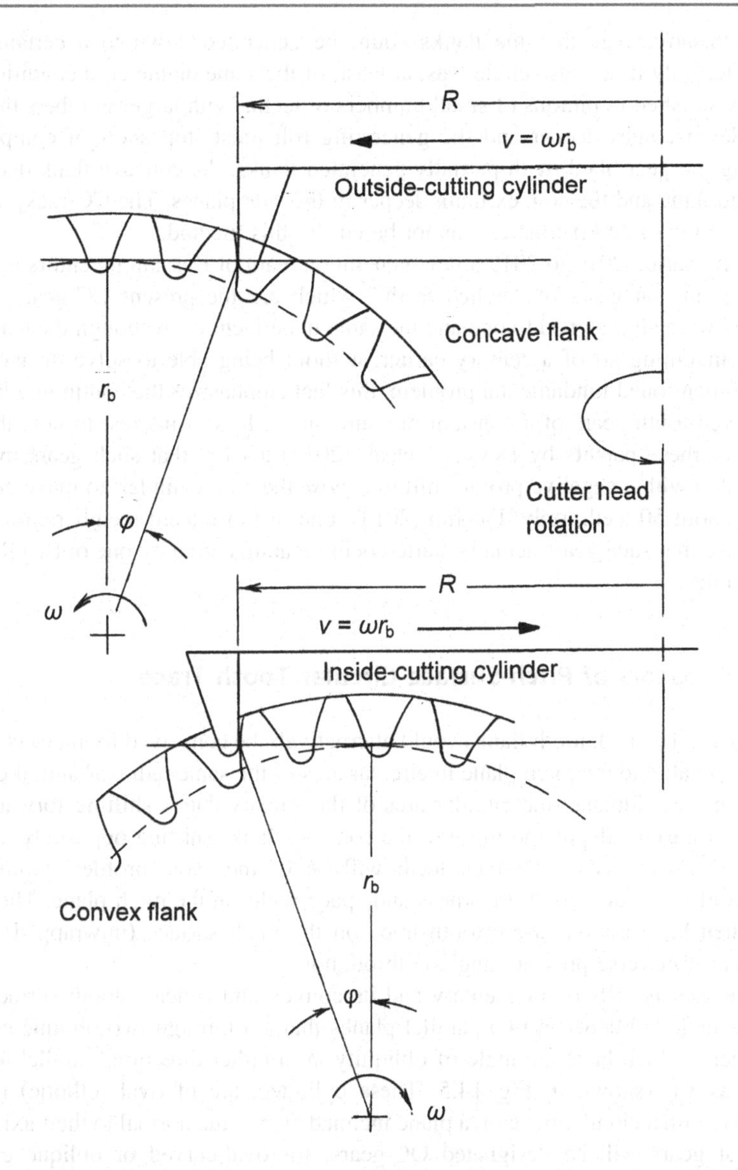

Fig. 14.6 Generating kinematics of curved-tooth gears having a circular tooth trace on the base cylinder, drawn in the instantaneous position of the blade cutting edge in the gear midplane with its single cutting point on the pitch circle

done only by the circular edge of the cutting cylinder, before any nose rounding, which edge must be fed along the plane of action; tangent to the base cylinder. The rounded nose of the blades would then generate the root fillet; its lowermost locus being fed in a plane tangent to the root cylinder. Therefore, these gears suffer a major disadvantage that the flanks could be generated down to a certain form diameter only if the base circle was, at least, of the same diameter, a condition that is only satisfied in pinions of small numbers of teeth. With larger numbers the base cylinder becomes deeper and the generating roll must stop short of completion, leaving the gear blank with partially generated flanks, the concave flank deeper in the midplane and the convex flank deeper in the side planes. The CC-rack, used to describe the basic kinematics, cannot be cut by this method.

In the period 2010 to 2012 a new-generation team of Russian inventors received five patents on gears of "arched teeth," which are the present CC-gears. These patents were all assigned to one and the same manufacturer. Although the team was still reinventing art of a century earlier, without being able to solve or avoid the above-mentioned fundamental problem, this fact emphasizes the continuing interest in curved-tooth gears of a constant pressure angle. It is of interest to note that the latest of these patents by Davydov et al. (2012) teaches that such gears must be generated with a negative profile shift to *expose* the base cylinder; to make gears of up to about 50 teeth, only. Parshin (2014), one of those team members, furnished evidence that such gears actually started being manufactured by one or two Russian companies.

14.4.5 Gears of Pitch-Surface Circular Tooth Trace

OC-gears The rack tooth flanks could alternatively be conceived to intersect all the planes parallel to the pitch plane in circular arcs of the same radius R and of centers lying in the midplane; the circular arcs of the convex flanks shifting forwardly in proportion to the depth and those of the concave flanks shifting oppositely. All the transverse sections of the rack teeth will be of the same profile; symmetrical trapezoid with equal tooth thickness and space width in the pitch plane. The gears will then have a circular-arc tooth trace on the pitch surface (unwrapped) and a constant transverse pressure angle φ throughout.

The rack is self-complementary and its convex and concave tooth surfaces are stripes included between two parallel planes that cut through two oblique circular cylinders, which have an angle of obliquity φ in either direction, parallel to their base, as was shown in Fig. 14.5. These cylinders are of oval (elliptic) normal section, with a circular trace on a plane inclined at φ to the normal to their axis. This type of gears will be designated OC-gears; for oval-curved or oblique circular cylindrical rack tooth flanks. OC-gears of finite numbers of teeth will then have convex and concave flanks of which each fiber in the gear transverse plane is an involute that stems off the base cylinder. In mesh, OC-gears will be in full-face line contact which moves in the plane of action.

It is of interest to find out that this type of gearing had already been suggested in a long-forgotten patent by Letzkus (1868). In this document the description of the geometry—that should have been based on the rack—seems to take for granted that the gear and rack are conjugate. But the invention did not suggest any cutting process.

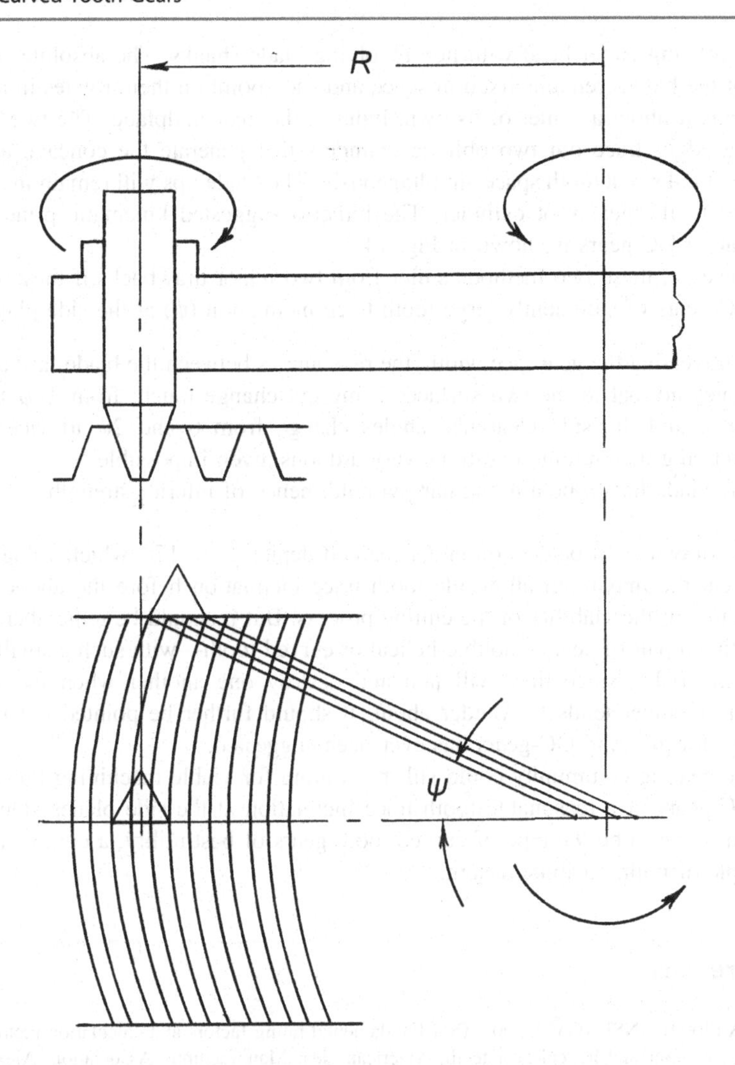

Fig. 14.7 Kinematic principle of generating curved-tooth gears having a circular tooth trace on the pitch cylinder

Generating-cutting OC-gears does not seem to have received due attention; it is a difficult task indeed. Two methods have been suggested over the years, which are based on one and the same kinematic principle. This principle envisages making each of two oppositely inclined, straight cutting edges on a non-rotating tool rake surface perform an oblique circular motion parallel to itself. The two edges could thus be of one V-shaped blade, and a set of such blades could be used to act either in parallel, viz. a rack cutter in reciprocating, arcuate parallel motion, or in series, viz. a rotating cutter head with non-revolving blade shanks. The absolute orientation of the blades remains fixed in space and each point on them rotates in a circle of radius R about a center of its own, lying in the gear midplane. The two sets of cutting edges trace out two oblique cylinders that generate the concave and the convex flanks of a tooth space simultaneously. The blade tips will remain in a plane tangent to the gear root cylinder. The hitherto suggested kinematic principle of generating OC-gears is shown in Fig. 14.7.

However, these two methods suffer from two major drawbacks if they were to cut OC-gears of sufficiently large tooth trace inclination (ψ) at the side planes:

1. In traversing the gear face width, the rake angles between the blade face and the planes normal to the two surfaces being cut change much; from $\pm\,\psi$ to vice versa, and the side clearance angles change from 0 and 2ψ to vice versa, rendering the cutting conditions very arduous, even impossible.
2. The blade has to be a triangular pyramid, hence of inferior strength.

The drawings in patents on either method depict $\psi\ \leq\ 17°$, which value might represent the maximum allowable tooth trace inclination before the above drawbacks inhibit the viability of the cutting process. But it should be remembered that operating a pair of gears—double-helical or curved-tooth—with such a small angle is highly risky, since they will jam and destroy one another when the axially floating member tends to wander about. It should further be pointed out that no process for grinding OC-gears has ever been suggested.

The gearing community could still be waiting for viable machining kinematics for OC-gears of a reasonable tooth trace inclination at the side planes since they should prove to be *the* type of curved-tooth gears of best utility; a comprehensive example of multi-attribute design.

References

AGMA (2004) ANSI/AGMA 2001–D04 Fundamental rating factors and calculation methods for involute spur and helical gear teeth. American Gear Manufacturers Association, Alexandria, Virginia

Andrei G (2009) Meeting component reliability needs of the aerospace industry. Paper presented at the reliability seminar of Avio S.p.A., Aarhus, Denmark, 6 October 2009

Arafa HA (2005) C-gears: geometry and machining. Proc I Mech E 219 Part C J Mech Eng Sci (7):709–726. https://doi.org/10.1243/095440605X31481

Arafa HA, Bedewy M (2010a) Quasi-exact-constraint design of wind turbine gearing. In: Proceedings of the ASME 2010 power conference, Chicago, Illinois,, 13–15 July 2010, pp 607–616. https://doi.org/10.1115/POWER2010-27012

Arafa HA, Bedewy M (2010b) C-gears in rotorcraft transmissions: a novel design paradigm. In: Proceedings of the AHS/AIAA/SAE/RAeS international powered lift conference, Philadelphia, Pennsylvania, 5–7 October 2010, pp 93–100. Curran, New York

Bittner EH (2018) Machine for machining gear teeth and gear teeth machining method. US Patent 9,956,631, 1 May 2018

Böttcher P (1913) Bogenförmiger Zahn für Stirn- und Kegelräder. DE Patent 265809, 16 October 1913

Davydov AP et al (2012) Arched cylindrical gear transmission. RU Patent 2469230, 10 December 2012

Letzkus J (1868) Improvement in teeth for gear-wheels. US Patent 80,291, 28 July 1868

Lewicki DG et al (2008) Face gear surface durability investigations. J Am Helicop Soc 53(3):282–289. https://doi.org/10.4050/JAHS.53.282

Parshin AN (2014) Arched toothed cylindrical gears manufacture on CNC lathes and experience of their inculcation. Proceedings of the international symposium theory and practice of gearing—2014, Izhevsk, Russia, 21–23 January, pp 151–159

Printed in the United States
by Baker & Taylor Publisher Services